WHY CAN'T WE BE MORE LIKE TREES?

"Judith Polich has given us a magnificent gift in writing *Why Can't We Be More Like Trees?* She brilliantly points out that we must awaken to the ancient indigenous wisdom that everything is alive, is conscious, and interconnected. In *Why Can't We Be More Like Trees?* she inspires a more holistic approach to life through her emerging narratives and gets readers to think outside the box. Judith is so passionate about waking people to an entire new level of consciousness. A remarkable book written so beautifully, it has a healing energy that can be felt through every page."

SANDRA INGERMAN, M.A.,
INTERNATIONAL SHAMANIC TEACHER,
COAUTHOR OF *SPEAKING WITH NATURE,*
AND AUTHOR OF *WALKING IN LIGHT*

WHY CAN'T WE BE MORE LIKE TREES?

The Ancient Masters of Cooperation, Kindness, and Healing

JUDITH POLICH

Bear & Company
Rochester, Vermont

Bear & Company
One Park Street
Rochester, Vermont 05767
www.BearandCompanyBooks.com

Bear & Company is a division of Inner Traditions International

Cataloging-in-Publication Data for this title is available from the Library of Congress

ISBN 978-1-59143-504-4 (print)
ISBN 978-1-59143-505-1 (ebook)

Printed and bound in the United States by Lake Book Manufacturing, LLC

10 9 8 7 6 5 4 3 2 1

Text design and layout by Kenleigh Manseau
This book was typeset in Garamond Premier Pro with Gala and Avenier Next used as display typefaces

To send correspondence to the author of this book, mail a first-class letter to the author c/o Inner Traditions • Bear & Company, One Park Street, Rochester, VT 05767, and we will forward the communication.

Contents

Acknowledgments

My gratitude goes out to my many thoughtful readers and friends, Marty Tombari, Karen Peterson, Darius Strickland, Joe Hardy, Alice Schleiderer, Shirley Freriks, Ginny Stearns, Virginia Reed, and Kristin Clausen, whose comments helped me refine and develop the concepts found in this book. The insight and assistance provided by my friend Jane Thimke deserves special acknowledgement. I am especially grateful to my partner, Gayle Dawn Price, for her support, encouragement, and technical assistance. I am also grateful to Nick Mays, who loves trees as much as I do, for his editorial assistance.

Author's Note

This book is about narratives—the new and old stories that shape our behavior. We examine the tired stories that have brought us to the edge of a human-created apocalypse and the emerging stories that, with luck, will keep us from falling over the brink and may even lead us to the next phase of planetary evolution and a greener, more conscious world.

INTRODUCTION

> *Oak* (Quercus)
> *A large long lived, deciduous tree native to Europe, the Caucasus, Asia Minor, and North Africa. . . . Together with a long list of economic benefits it bestows, the oak seems to have a rather caring quality. It is not surprising that the ancient Gauls and the Romans associated the oak with Mars Silcanus, the god of agriculture and healing.*
>
> THE MEANING OF TREES

My earliest memories are about trees. I grew up on a farm in northern Wisconsin near Lake Superior and within hollering distance of a million acres of national forest. There was a small grove of Norway spruce in our backyard. It functioned as my playground. It contained a small sandbox and a home for all my outdoor toys. I could barely walk, but I remember crawling around on the spongy, fragrant forest floor layered with soft red-gold needles. The canopy provided wonderful shelter from sun and wind. It smelled tantalizing, pungent, piney, and, in a way, delicious. Sometimes I would lie on my back and timelessly watch the upper branches sway gently in the breeze. Sometimes I would fall asleep, dreaming my own version of "rock-a-bye baby on the tree top." I always felt safe and wonderful there—a small, solitary child, almost hidden under the large and fragrant interlocking branches of what was a mother tree and six of her offspring.

As soon as I was old enough to wander off by myself, I spent most afternoons, when I was not in school, in the nearby woods. Our farm bordered a pastured woodlot owned by my aunt. It was a jungle of poplar, maple, hazelnut, raspberry bushes, and an occasional oak. My hangout was what I thought of as a giant oak about a quarter of a mile from our house. After crossing a little creek and dipping under the barbed-wire fence, I was off to be with my friends—my tree friends. I did not consciously think of them as tree friends, but they were all of that and more. It was just my place, and I knew I was always welcome and protected there. And, of course, it was magical, filled with little creatures and smells and adventures. I would lean up against my oak, watch the poplars quaking in the breeze, take in the smells of the season, hunt for raspberries and hazelnuts, and listen to the red-winged blackbirds calling from the cattails and the loud and joyful frog chorus in the nearby pond. There was no place I would rather be. Mostly it was just peaceful—a place I always felt calm and connected. What I felt connected to was everything around me: the grass, the trees, the sky, the birds—everything. It was all vibrantly alive, and I was a part of it. I did not know it at the time, but this was not normal. I did know enough not to talk much about it though. . . .

Not far from my oak were several giant blackened tree stumps. They were as wide across as I was tall. Their hollowed-out centers made great natural thrones filled with moss and treasures. I would climb inside, place myself on these thrones, and absorb the marvels of my forest kingdom. I would later find out that they were the remnants of a massive forest of mature white pine that had covered much of the north-central and northeastern tier of America. These old-growth forests were logged out in the eighteenth, nineteenth, and early twentieth centuries. Subsequently, their stumps were blackened by fires. Some trees were over five hundred years old. They grew up to 230 feet tall.

Their canopies were so thick and enclosed that it was said that squirrels could jump from treetop to treetop for miles.

Occasionally, when I was older, I would go back to visit my oak. It turned out to be not nearly as big as I remembered. It all seemed ordinary—just an average-sized oak at the edge of a pasture near a small, muddy pond. The magic was gone. I had lost the ability to see with an unconditioned lens.

When I was in my twenties, I fell back in love with the natural world. My eyes again opened to the wonders of the world around me. I once again became a creature of the forest. And I was not alone. This was the beginning of the environmental movement. My tribe and I lived in and loved in the forest. And we felt loved in return.

As a graduate student in the newly formed Institute for Environmental Studies at the University of Wisconsin, Madison, I would take undergrads out to the woods, on the rivers, and tromp around in the spectacular valleys and glens of southern Wisconsin, where I would let nature's magic do its work. It was called immersing. Sometimes we would bury ourselves in the northern bogs, as Aldo Leopold had, and just lie there, our noses above water, soaking in the marsh world. And, of course, we would swing high in trees in the wind, as John Muir had taught us.

Immanent immersion and God were everywhere. Awake to the living reality all around us. And then I went to law school. As it happens, I took a turn in the road. Law school, like any professional training, is also an immersion—an immersion into a narrow, precise worldview. It requires adherence to a linear and compartmentalized conceptual understanding of norms and rules and codified consensual realities. It uniquely hones the development of the rational, analytical part of our mind, at the cost of our more holistic, inclusive attributes.

I was a good student and a good lawyer, but I was inwardly conflicted. There was really no such thing as a holistic lawyer, but I did try again and again to approach my professional life with some sense of integration. Ultimately, I had to split my worlds. There was work and its heady, but dull, abstract complexity. There was the rest of my life—the real part, which I considered my spiritual self—with its wild adventures with shamans, power points, Inca masters, and gurus of all persuasions. But I came to understand that what I truly sought was to commune directly with nature—with all of life. At some point, I realized that for most of my adult life, I had been trying to reclaim the innocent acceptance of the vital interconnection with the whole of life that I felt as a child in the nurturing embrace of my old friend, the oak tree.

I came to understand that duality was merely a false convention. I realized that my personal journey of integration and reconciliation of my rational ego-based self and my spiritual, holistic, nature-based self was my path. I understand now that my process is really everyone's journey, as we all stand now at another turn in the road.

ABOUT THIS BOOK

Chapter One, "The Heart-Brain of the Forest," tells the story of the forest and plant networks based on revolutionary research that uses new technologies to allow us to "see" the extraordinary interactions and communication that happen underground, in a world we never before thought of as sentient, much less a world from which we can draw great wisdom. We discuss how trees communicate, how they share resources and care for their offspring, the elderly, and the infirm, all the while expressing behaviors that can only be described as cooperative, altruistic, and holistic.

Chapter Two, "Finding Our Place in Nature" describes how plants, throughout our evolutionary history, created diverse ecosystems that animals thrive in. It then tells us the story of how humans—and all animals, since we do not photosynthesize—have been shaped by a very different evolutionary pathway, which requires different and more competitive strategies for survival. We explore how the path of predation led to the development of a central brain and nervous system and hardwired instincts. We discuss the cognitive revolution, which apparently was due to a random mutation in our species, *Homo sapiens,* and led suddenly to new mental abilities and perceptions, including abstract thought, reductionism, the subject-object dichotomy, and other mental constructions that led to false narratives and that placed us outside of nature. This radical shift led to our advance as the dominant species and ultimately to our disastrous impact on our planetary home.

Chapter Three, "How Nature Heals Us," describes how the plant kingdom, and specifically the forest community, offers us many avenues to heal our sense of dissociation and disenchantment, as well as correct the many misconceptions we harbor that place us outside of nature. We explore new narratives emerging in science that explain how spending time in nature de-stresses us, calms the limbic system, enhances our immune system, and increases our serotonin levels. We explore the history of healing plants and our apparently innate biophilia: our love of nature and the natural world. We discuss how time in nature helps our overstressed minds rest and relax, and the role neurohormones like oxytocin may play in this process. Finally, we explore what new neurological research suggests about how nature helps us drop our sense of separate self and shift to a more holistic perspective by suppressing our brain's default-mode network, analogous to what happens when we meditate or take mind-altering substances like LSD and psilocybin.

Chapter Four, "Our Tree Connections," examines the attributes we share with plant and forest communities and how we are radically different from each other. We may be competitive, but we are also the most social and cooperative of all mammals. Social science tells us that cooperation, altruism, and kindness are now believed to be genetically coded in humans. Emerging narratives suggest that natural selection may favor mutual and communal benefit rather than individual benefit. We discuss the role of neurohormones like oxytocin and cannabinoids in the development of these behaviors. Plants, we find, also produce neurohormones. We explore the results of research on plant neurobiology and plant neurochemistry and the extraordinary dispersed and collective intelligence and sensory awareness of our green allies. These new narratives point to major perceptual shifts.

Chapter Five, "Greening Our Stories," examines how our spiritual belief systems and practices are slowly greening as we face a climate catastrophe. We examine how the evolving narratives found within our four major religions, Christianity, Islam, Hinduism, and Buddhism, which are adhered to by some 6.5 billion people, are greening and encouraging sane practices and policies like sustainability and integral ecology. Pope Francis's extraordinary 2015 treatise, *Laudato si'* (Praise Be to You), clarifies the Christian story of creation and our role as custodians and explains that there is no justification for our exploitation of nature. Islam offers its view as a model for sustainable development and green policy based on the Qur'an. Hinduism tells us again that everything is alive, conscious, and inseparable. Buddhism offers mind-training practices to help us deconstruct harmful mental conditioning and biases and develop greener behaviors. That is the overstory. The understory shows us that our older animistic beliefs are still with us and that our religious stories may have in fact originated from mysterious and surprising plant-based sources.

Chapter Six, "Seeing with a Greener, More Humble Lens," discusses dangerous cognitive biases such as plant blindness and mindless consumerism. We explore the many ways in which trees and all plants make our lives possible. In addition to providing almost all our food and most of our materials, our green allies filter the air, provide oxygen, remove CO_2, create soil, and prevent soil erosion. They also play a critical role in the global water cycle. We discuss how, with the development of widespread agriculture, our brains may have actually changed and learned to compartmentalize and think in a linear way. Our perception became dualistic and fragmented, filled with anthropocentric self-importance. We are now embracing new stories coming from both science and religion. This new worldview is based on concepts like sustainability, integral ecology, and a reconsideration of anthropocentrism. This new narrative is telling us that we cannot understand the whole by simply studying the parts; we are part of an interconnected complex network. Science is now embracing emerging theories of consciousness that tell us we are part of the community of life, and that our old narratives no longer have survival value.

Chapter Seven, "Restoring, Rebalancing, Regreening," considers just what our role as steward and custodian means as we try to stop the wave of planetary destruction we have set in motion. Our planet is on fire, we are reaching climate tipping points, and the Earth is moving toward the sixth extinction. But there is still time to reverse course if we act on the emerging holistic narratives and direct our unique brilliance and innovative nature to the critical transformation essential for our survival. We examine how the loss of trees and the destruction of habitats can be reversed by tree-planting, restoration, rewilding, and reclaiming land and waters we have destroyed. Trees, which are truly the planet's super species, may well be the guiding archetype in this massive effort to rebalance our planetary home. This transformation involves a new contract with the other species with whom we share the planet. That means we must live up to our responsibilities—the obligations—that came with the key to this amazing garden of life we call home. We are offered a glimpse of what the new blue-green world we are co-creating with our green allies will look like.

Finally, I would like to point out that I am not a scientist, historian, or philosopher. I am just a curious and optimistic person who loves nature.

1 The Heart-Brain of the Forest

Birch (Betula)

Named for the whiteness of its bark, the birch shares its name with the ancient Irish goddess Brigid . . . a benevolent deity, a muse to poets and patrons of crafts, particularly spinning and weaving. In Norse and Germanic tradition, the birch is associated with Freya, the Lady of The Forest. . . . The nourishing and caring birch is an image of the White Goddess . . . and stands for motherhood, bosom, and protection.

The Meaning of Trees

A new story is beginning to surface, based on revolutionary new technologies that allow us to "see" the extraordinary interactions and communication that happen underground. These techniques reveal a world we never before thought of as sentient—much less a world from which we can draw great wisdom. We are learning to see with new eyes, using new modalities, to extend our very limited senses. In the last couple of decades, research conducted in the soil under trees and other plants has completely revolutionized how we think about trees. This research has been made possible by the availability of state-of-the-art molecular and genetic tools, which allow for low-cost DNA sequencing and advanced molecular microscopy. Scientists can also now make use of tools like mass spectrometers to measure minute amounts of light in underground worlds at molecular levels. What we have learned will amaze you.

Molecular imaging allows the visualization of cellular functions in real time, without disturbing the organisms being studied. Probes known as biomarkers are used to help image pathways or targets; the biomarkers interact chemically with their surroundings and alter an image based on the molecular changes that occur in the area of interest. This technology allows for the imaging of very fine molecular changes. DNA sequencing is now used in molecular biology to study genomes and the proteins they encode. High-throughput sequencing technologies are capable of sequencing many DNA molecules in parallel to create large data sets. Mass spectrometry measures the mass-to-charge ratio of ions. Scintillation counters detect and measure ionizing radiation.* These technologies allow us to see into and analyze the complex dynamics of plant communities and other "alien" underground worlds we could previously only speculate about.

Although these tools were not designed to study soil, as the costs of their use decreased, they have become widely used in many applications and have widened our understanding of formerly hidden worlds. Together these tools have allowed scientists to "see" precisely what happens deep below the forest floor at a cellular level. With this technology, scientists have determined not only that plants communicate—but how they communicate and cooperate. We now know that complex signaling pathways (communication networks) exist between tree roots and fungi, and we can observe the cellular processes that underlie these

*See Merriam-Webster for definitions of mass spectrometry, DNA sequencing, molecular imaging, scintillation counter, and other technologies mentioned.

relationships. Scientists have been able to actually observe the otherwise invisible underground exchanges of carbon sugars and minerals being transferred back and forth between trees at the root level.

Before these new tools were available, we could study the soil with a magnifying glass, but we could not see much other than worms, snails, centipedes, and other barely visible creatures. Most of the creatures that live in the soil are far too small to be seen, even with magnification. Although we cannot see many of them without advanced technology, the soil is full of living creatures. They include archaea, bacteria, actinomycetes, fungi, algae, and protozoa, as well as the larger animals we can see visually, like earthworms, insects, and burrowing mammals. Many of the insects living in the soil are arthropods and have exoskeletons. They include mites, millipedes, centipedes, springtails, and grubs. Up to ninety-five percent of the biomass produced by green plants is consumed by microorganisms. Fungi, in particular, are critically important and make up a large component of the soil community.[1]

There is still a lot we don't know about soil and its complex lifeforms. Most have yet to be identified. We do know that half of the biomass of a forest is below the ground. It has been said that there are more life forms in a handful of forest soil than there are people on the planet. They say that each teaspoon of forest soil contains miles of fungal thread. And what that fungal network does will amaze you.[2]

Previously, we had no idea what happened in the soil at the root level of a plant community. We had no idea how complex life is down there. We were only focused on what we saw above the ground. Trees were either attractive, useful, or a nuisance. They provided shade and added interest to our landscapes. Mostly, they were objects, commodities used as fuel, for lumber, for making paper, and for creating many of the products that make our lives better.

But in the last thirty years, that has changed. Scientists now know a lot more about trees and the forests they live in. With their new tools, they have watched and recorded the exchange of information, revealing a complexity that has forced us to rethink our basic assumptions. They have found the many ways in which a plant community communicates, the degree to which information is widespread within a forest or ecological network, and the elaborate way in which the forest community functions. This research, which has only recently been popularized for the nonscientific reader, is forcing us to rethink everything we know about trees, forests, and plants.

There are now many researchers around the world who are using these new tools to look into the microbial world of soil and microbes and to shed light on the complex social network that exists in an established forest. It no longer makes sense to think of a tree as an individual plant—a tree is an interactive part of a cooperative network. The complexity of that network raises issues of whether trees have consciousness, what kind of consciousness that might be, and whether trees are, in fact, sentient. It is fair to say that a new paradigm is arising in the world of plant ecology. Our long-held assumptions are on very shaky ground.

This all started in 1992, when a researcher named Suzanne Simard was working on a doctoral thesis in the Pacific Northwest. She was studying a plantation and found that when the paper birch trees were cut, the Douglas firs suffered and died prematurely. Of course, that is contrary to what we believed would happen. Foresters thought that getting rid of the birch would have left the firs more sunlight and capacity for photosynthesis. Using new technological tools like mass spectrometers and scintillation counters to study what was going on well below the surface, she stumbled on something unexpected. Using radioactive isotopes in carbon, she found that a huge below-ground network existed between the tree roots of the birch and the fir and symbiotic fungus. We now know, based on the fossil record, that the association between fungi and plants is an ancient one.

Mycelium is the vegetative part of a fungus and consists of branches of thread-like hyphae, which are tiny branching filaments. They are made up of cells that include nuclei and hold genetic material. Their job is to absorb nutrients and transport them to other parts of the fungus. Collectively, they are called mycelium and can run for miles deep underground.

Simard found that deep in the forest floor, there was a huge web of tiny mycelial (fungal) threads that interconnected with the roots of the firs and the roots of the birches, as well as the entire forest network. It was an interactive network. Using the new technology that makes the molecular level visible, she was able to see that the extensive mycelial network transferred carbon—that is, photosynthetic sugar—from the birches to the firs, enhancing the growth of the firs. She found that in some seasons, the firs returned the favor, feeding their forest friends. This was not a forest of individual trees and species; it was a network that cooperated and shared resources. It was a single whole. And, as a single whole, it had internet-like, complex communication. She called the communication mechanism facilitated by the mycelial fibers "the wood-wide web."[3]

In a recent book titled *The Hidden Life of Trees,* German forester Peter Wohlleben tells the story of the life of the forest based on similar research conducted around the world and, in particular, in the forests of Germany.[4] He explains in detail what can only be described as true tree friendships. Trees live a long time—hundreds of years—but their lives are very slow compared to ours. They will stand in a forest for their entire life with their roots intertwined with those of neighboring trees. These are tree friends. They communicate with each other, they share food with each other, and they defend each other. These trees live within a rich social network, a network based on sharing, cooperating, and communicating. Not only do they nourish the neighboring tree friends of the same species, they also nourish trees of different species.

Wohlleben explains that there are many advantages to working together. Together, the trees can create an ecosystem that controls temperature extremes and retains more moisture. In an apparently thoughtful and precise manner, they regulate their growth and enhance the growth of others to jointly create a protected environment. While it's true that trees will compete for sunlight, their care for the community transcends any individual competitiveness. They are careful not to grow thick branches in the direction of a neighboring tree. They don't take anything more than they can utilize from their other tree friends, and, in fact, will transfer food and nutrients to their tree friends. They will assist if their friends are ill by significantly supplementing their nutrition. Sometimes tree friends even die together. Most of the work Wohlleben cites has been done in undisturbed forests, and he is quick to point out that this isn't always the case in tree plantations, where trees are more isolated and suffer more, accordingly.

Research conducted at the Environmental Research Institute in Aachen, Germany has shown the degree to which trees will share resources with others.[5] Trees grow in different natural conditions. Some have more water in their soil. Some grow in rocky soil. Some are found in rich humus, high in nutrients. Logically, some trees should be able to photosynthesize better than other trees. But according to this research, they all photosynthesize at an equal rate, regardless of their underlying conditions. That can only be possible if they are sharing resources under the ground so that they all can produce the same amount of sugar per leaf. This research has shown that this sharing is conducted at the root level—whoever has excess sugar transfers it to those who are lacking through their roots. I find this absolutely fascinating, as it obviously also implies a certain amount of computational skill. The research also shows that in some species, it is more productive for trees to grow close together so it's easier for them to help each other.

The Darwinian principle of survival of the fittest does not seem to apply to trees. It's pretty clear from the research done in the last thirty years or so that the well-being of a tree depends on the well-being of the forest community. Balance, harmony, and cooperation are the keys to survival.

Walking in forests is my favorite pastime. I never grow tired of the same trails. I love watching the subtle changes week to week and the dramatic changes season to season. Yet, my forest walks have changed since I learned of this research. I have realized that I saw only a fragment of what exists, and I misunderstood what I was seeing. I now think of it like this: I am walking in the forest and I see a stranger coming toward me, but I only can see the top half of his torso. That is what I expected to see because I did not realize there was more to him. Now I understand that this person coming toward me has a full body, just like I do, with arms, legs, and feet that touch the ground; furthermore, he is not alone. He is a part of a large interactive community. They are all tree friends, chatting with each and sharing insights. I want to hear what they are saying. I want to get to know them more intimately. Now, when I see a tree, I know there is far more to that tree than what I see at the surface. Of course, there are tree roots that extend deep into the soil. But there is more. There is an entire community that is exceedingly complex and interactive that I can't see, but that I know exists. This new knowledge informs my limited perception. A walk in the forest becomes a much deeper exploration.

Wohlleben also tells us that research has shown that trees feed and nourish their offspring. And yes, research shows that they can distinguish their young from other saplings. Other research shows that trees send nourishment to a tree that is ill or injured. That would seem to be an act of compassion or kindness. For what other reasons would they give away precious resources? They also take care of the elderly, supplementing the nutrition and water needs of older trees so they can continue to grow. This seems counterintuitive; older trees take up more of the canopy and therefore get more sunlight. And, as an additional altruistic act, trees even transfer their nutrients to neighboring plants before they die.[6]

One of the most fascinating stories Wohlleben relates is his discovery that vestiges of the stump of a tree that died 400 to 500 years ago showed signs of chlorophyll, indicating that its roots were still alive. Those old roots could not photosynthesize themselves, of course, but they were apparently being fed by the underground tree root and fungal system of the forest network. Why would they do that?

These are mind-boggling findings. Why does the forest network work in this manner? Why do they feed and protect the ill, elderly, infirm, and the young who cannot care for themselves? How is this possible? How do they communicate such detailed, precise, even mathematical information? Does this mean they have cognition? It suggests some level of consciousness. Obviously, these findings have turned the scientific community on its head.

Let's start with communication. There are numerous ways plants communicate. They obviously don't speak in the manner we do, but they do seem to have a language. It's been documented that one form of their language is through smell—or rather, through the scents that they emit. For example, when an acacia tree is being attacked by insects, it will emit a gas, like ethylene, into the air to warn other trees. Those trees will perceive the ethylene and react in turn by creating specific toxins in their leaves that taste bitter to the predators. The scents released vary with the type of attack. It's been demonstrated that trees will release scent compounds that identify specific types of predators. That suggests an ability to analyze in some manner. These different scents convey specific information. They are like words, understood by both sender and recipient. This scent language tells other trees what kind of toxin would be effective. And their scent vocabulary in extensive. Trees can also release a pheromone that will attract a predator of the bug or insect that is attacking it. Their scent language seems to have many words, with specific meanings. In some cases, the scent tells other trees to create a specific toxin. Other scent messages are even more complex, and they direct a tree to create a specific hormone or pheromone. In *The Biophilia Effect,* Charles Arvay analyzes Swiss biologist Florianne Koechlin's findings where she reports that over 2,000 plant fragrances or scents have been found from over 900 plant families. She refers to them as chemical words and says that most are organic compounds called terpenes. That seems like pretty sophisticated language, and it is only one of the ways plants communicate.[7] These findings raise a lot of questions about plant intelligence. Significantly, plant behaviors do not seem to be instinct-driven responses. Obviously, if we can even consider applying the term intelligence to plants, we must first acknowledge that their form of intelligence is very different than ours and seems alien to us. But we must ask how this is all possible. And, ultimately, how they got so smart.

I realize I am asking you to stretch a bit here when I suggest that plants have intelligence, are smart, can analyze, and may be conscious. Bear with me. More will be revealed, and you can form your own conclusions. Just keep an open mind. And remember that not all that long ago we did not believe that animals were intelligent or that they had feelings. We did not think they had consciousness or that they could experience pain and suffering. We now understand those premises were false.

We have learned that trees also communicate through their fungal networks using chemical signals and electric currents. There are huge fungal networks that connect individual tree root systems to the root systems of other trees throughout the forest. Those fungal networks broadcast information to all the trees in the network so the trees in the network can act to defend themselves. These fungal networks can cover miles and transmit signals from one tree to another, exchanging information about insects, drought, and other dangers. Trees that have a wood-wide web lead healthier lives and have increased abilities to defend themselves as well as to equitably distribute resources within their community.

Researchers have also found that tree roots send out and receive clicking sounds, perceivable as ultrasonic frequencies that can be heard by other trees. They may do so in times of stress, such as when they are running out of water. They may do so to alert other trees to the need to be more conservative in their water use. They may use ultrasonic frequencies to communicate other matters. This behavior is similar to echolocation. This emerging research field is called bioacoustics, an area that scientists are only beginning to investigate. Research like Monica Gagliano's also explains that plants communicate by bouncing light signals off their neighbors.[8]

It is estimated that about ninety percent of all trees and plants have underground symbiotic fungal networks. These fungi or mycelial networks, which are invisible to us, are massive. Some can cover hundreds of acres. Not all fungi are cooperative or beneficial. But overall, they have evolved with their green hosts in a positive manner, and it works for both of them. Vital nutrients are passed through the fungal membranes. The membranes retrieve minerals from deep in the soil, where these nutrients would otherwise be out of a tree's reach. These are nutrients like nitrogen, phosphorus, carbon, and magnesium, which are essential for the tree's growth. Fungi can be surprisingly proactive on behalf of their tree allies. One of my favorite stories that was microscopically observed is of mycelial fibers from an oyster mushroom. Like cowboys, using their fibers, they corralled, lassoed, and strangled nematodes living in the soil and carried off these nitrogen-rich critters, transferring their vital nitrogen load directly to the roots of their tree allies. These fungi hunt and mine on behalf of their partners. That is surely something a tree root would be incapable of doing on its own.

The fungi, which are neither plants nor animals, cannot photosynthesize. Instead, they have developed an ingenious arrangement to secure their nutritional needs. They send out mycelium threads to explore their environment and pick up nutrients like phosphorus and nitrogen, as well as water, and bring them back to their plant buddies. The plant trades their mycelial partner carbon sugar for the nutrients and other things it needs. You could say it is a simple "root level" form of capitalism. The fungal system connects one set of tree roots to another so that carbon, water, and items needed for defenses can be exchanged between trees as well. The mycelial network and the trees work together as symbiotic partners. There is some debate about whether the relationship is like a business deal or purely symbiotic. Fungi are very smart and should not be underestimated.[9]

Trees get the information they need from the wood-wide web. They also benefit from the essential nutrients and the specific defenses the fungi pharmacy produces for them. In exchange, the fungal networks get food. Their fungal partners will also filter out many toxins and heavy metals that are detrimental to the trees. These are just some of the interactions we know about. Apparently, though, fungi drive a hard bargain. Research has indicated that trees transfer as much as one-third of their photosynthetic sugar to their fungal partners. Some research indicates they take far more. And they also act as banks, storing sugars the trees and plants may need later. It is a truly functional partnership.[10]

Some symbiotic fungal networks and tree partners live, work, and dine together for hundreds of years. Some fungi only associate with a particular tree species or set of trees. Yes, that is why those of us who hunt for specific mushrooms look around specific trees. Whatever the motivation, the symbiosis, even if mildly capitalistic, works for both parties.

I love mushroom season. The fungi fruits are the mushrooms we see above the ground. They are only a small part of the fungal network. Mushrooms are seldom solitary; they form in clumps, create circles known as fairy rings, and sometime seem to just magically appear. They tend to hang out with other fungi, many of which are toxic or—at best—not edible. I live in the mountains of the southwest. When our summers come, there is a migration of mushroom hunters that forage the higher peaks with great optimism. For weeks before, some say they dream of mushrooms. Sometime when foraging, it seems we are almost led toward the mushrooms we are seeking. It is as if they entice us by beckoning us to climb high in the mountains, gather their fruits, and distribute their spores. I think they may also have a type of intelligence. After all, they cut a pretty hard bargain with trees.

Simard and her students discovered what she calls mother trees. She and her students found that all trees in the forest community were linked and connected but that the oldest trees, the mother trees, were the most highly linked of all. Of course, bigger trees have larger root systems and a larger number of connections to the fungal networks. These older trees also bring a lot more carbon into the network and have more root tips. Through this work, Simard and others have been able to discover that these mother trees can recognize their own offspring and will continue to nurture their own kin for their entire life.

Simard uses terms like "forest wisdom" and "mother trees" to help us grasp the implications of her research. As humans, we can understand terms like "resource transfers," "kin recognition," and "defense signaling." She suggests that symbiosis between trees and fungi are like the neutral networks in human brains. It may be that the only way that we can truly understand these interactions is to think of them metaphorically, but we must be careful not to anthropomorphize. It is very hard to hear that mother trees feed their offspring and not anthropomorphize. It may be easier to think of them as like us but "other."

Plants are a lot more complex than we thought. They communicate with one another, with their fungal partners, and with animals. They communicate through every part of their body: their leaves and needles, their roots, branches, flowers, and their seeds. They communicate auditorily and through their other senses. They communicate in a manner similar to echolocation. They have memories. They can learn. We have to wonder: Are they conscious? Do they have a sense of self? These questions make a lot of traditional scientists uncomfortable.

Researchers have been able to halt plant motion by using anesthetics. The theory is that if plants can be sedated, then there must be some consciousness there that is being sedated; therefore, they must have some consciousness. However, this idea is widely disputed—often based on the logical idea that plants do not have a nervous system and brain.

Neuroscientists like Todd Feinberg and biologist Jon Mallet say plants do not pass the test for consciousness based on studies of insect and mammal brains.[11] Others, like Daniel Chamovitz, who was the director of the Manna Center for Plant Biosciences in Tel Aviv, insists that plants are neither conscious nor intelligent. He says they are complex, but he also states that we shouldn't confuse plant experience with human experience.[12]

The argument goes on and on, as it has for thousands of years. It goes all the way back to Aristotle's theory of biology, based on his personal observations some 2,500 years ago. He may have been the earliest to record writings on biological forms. It is important to note that these writings were based on his observations, not the scientific process as we now understand it. In *The History of Animals,* he arranges all beings in a fixed scale of perfection. We are on the top and plants and minerals are on the bottom. He did not consider plants as sentient but as a non-sentient animate material substance. And this archaic perception continued to the present. What it seems to come down to is that we define self and consciousness based on human selves and our ways of perceiving. That is anthropocentric, of course, but we cannot seem to get beyond our blinders.

The other side of the argument has been studied and well framed by behavioral ecologists such as Monica Gagliano. Gagliano and others have done pioneering research that determined that plants indeed can learn—that they have memory and a form of cognition. But to comprehend these implications, we need to let go of what Suzanne Simard calls "our anthropocentric superiority complex."[13] As we shall explore throughout this book, as humans we view the world through our unique sensory modalities, our lenses and our biases. Other species including trees have a different array of sensory modalities and different lenses. Our worldview may be unique to us, but it is not necessarily superior. We know that trees communicate in a number of ways. We know they have a sense of time. They know precisely when to drop their leaves. They know when to produce seeds. They know when to make new leaves to begin photosynthesis again. To accomplish these tasks efficiently, they must use some type of metabolic, mathematical process. But do they have cognition?

We know plants can perceive. They know when they have been attacked by insects and when they are injured. It would seem as though they can perceive pain. They immediately respond to an injury, whether it's an insect penetration, damage from a broken branch, or from pruning. They will send sap to seal their wounds in the same manner that we, when we are injured and bleed, will scab over our wounds. This prevents further infection and loss of vital fluids.

But *do* plants experience pain and have feelings? That is a bit more of a stretch. However, it turns out it depends on whom you ask. According to researchers at the Institute for Applied Physics at the University of Bonn in Germany, plants release gases that are the equivalent of screaming in pain. These researchers used laser-powered microphones to pick up sound waves we cannot hear that are produced by plants releasing gases when they are cut or injured. That may surprise most vegans. According to these researchers, these "plant voices" show that cucumbers scream when they are sick and flowers whine when their leaves are cut. If that is the case, imagine how a chainsaw feels to a tree. Similar research shows that plants can hear themselves being eaten and respond with defensive mechanisms.[14]

But we can relax a bit. Daniel Chamovitz explains this differently. He says trees may perceive a leaf being cut or an electric shock, but he stops short of saying they feel pain or suffer. He says that when we assume they feel pain or suffering, we are anthropomorphizing. They do struggle to regain homeostasis, but he says they do not have pain receptors like we do.[15]

I would like to point out that sometimes it is hard to know whether you are anthropomorphizing or whether you just don't have enough information to come to a valid conclusion. In other words, they may not have pain receptors like we do, but that doesn't mean they do not experience pain. As we shall see, this "other" we are learning about is quite complex.

We know trees are sensory aware. It's even been reported that plants can see us. They see us because they have photoreceptors that perceive different wavelengths of light. Chamovitz, the author of *What a Plant Knows,* says they can even distinguish the color shirt we're wearing. As will be explained in Chapter 4, they have all the senses we do. They appear to see, hear, taste, touch, and smell, and they have many senses that we do not have.

Gagliano states that since it has been documented that plants perceive, assess, learn, remember, resolve problems, make decisions, communicate with each other, and actively acquire information from their environment, those types of sophisticated behaviors indicate they have cognition. This is indeed a stretch, but her research and experiments on plants suggest that while plants don't have a central nervous system and brain, they behave like intelligent life forms. They seem to learn. They summon experience and knowledge repeatedly. They remember and learn from experiences.[16] Doesn't that imply a type of cognition?

The problem is that we define cognition from the only lens we have: our human perspective. We consider it a mental process or action of acquiring knowledge and understanding through thought, experience, and the senses. It involves thinking, problem-solving, and remembering—functions that are considered higher level capacities of the brain. These capacities can include language, imagination, perception, and planning. The obvious difficulty with plant cognition is that this human-centered definition seems to require a brain . . . or does it?

Our template for cognition calls for a brain and a neural system to actualize the complex processing involved in faculties like anticipation, awareness, memory, decision-making, learning, and communication. When human cognition is the reference point, interpretations of reality are experienced in terms of human values and perceptions. That is very limiting. But we simply can't seem to think of cognition in a way that is not anthropocentric. As Gagliano suggests, that limitation may be wired into our biology. In other words, our brains are wired to assign human attributes to others when they resemble human actions. Our brain understands our system of perception and movement but excludes other species that accomplish things in other ways. So, we cannot help but judge other species subjectively and deny their cognitive abilities if they use different modalities.[17]

Gagliano believes we need to approach this matter in a much wider manner. She cites the work of Chilean biologist Humberto Maturana, who suggests that organisms should be viewed as intrinsically part of the environmental niche in which they interact. As Maturana states in his book, *The Biology of Cognition:*

> Living systems are units of interactions; they exist in an ambience. From a purely biological point of view they cannot be understood independently of that part of the ambience with which they interact: the niche; nor can the niche be defined independently of the living system that specifies it.[18]

This approach is holistic. Cognition is viewed from a systems perspective. He explains that cognition is not a fixed property; it is not a thing in and of itself, but rather a dynamic process of interactions in the organism's environmental system. Cognition, then, is viewed as more of a process, as the organization of the actual functions and behaviors that make reactions possible and maintain production of further actions. That means cognition is a biological phenomenon contributing to the persistence of organisms in constantly changing environments. According to this theory, it makes sense to view cognition in humans, as well as nonhuman others and plants, as a functional process in the context of natural evolutionary relationships and their phylogenetic (evolutionary development) continuity. This approach does not require a brain and a nervous system. Nor is it clear whether the brain and nervous system in fact generate cognition. Maturana's approach also advocates for a paradigm capable of unifying a great diversity of expressions of raw cognition common to all living systems.[19]

We now know that animal behavior is more sophisticated than we ever thought. There is significant evidence of higher learning in animals, but plant research on this issue is not yet as advanced. Gagliano cites considerable research that's been conducted in the last ten or fifteen years on plant cognition, which reveals the extent to which a plant's perceptual awareness of environmental information directs behavioral expression. This research highlights that many of the behavioral feats and associated cognitive abilities exhibited by plants are easy to observe. Her most well-known study in this area is on the learning capacities of *Mimosa pudica,* which demonstrates the extraordinary perceptual awareness, learned behaviors, and memory of this seemingly lowly plant.[20]

Researchers in recent years have established that plants have the ability to assess relatedness and recognize and discriminate between kin and non-kin, both above and below the ground. Plants have a wide range of perceptual abilities. They know where they are in their environment in relationship to other organisms. They have a sense of place. They know where they are in their neighborhood. They are able to orient themselves based on cues like internal body centeredness, body posture, and external clues, just like we do. They respond to sounds and vibrations. In both plant and animal species, these abilities are important to avoid obstacles, find food, and outsmart predators.

Gagliano proposes exploring plant cognitive abilities in terms of the dynamic process of interactions in an organism's environment system, as suggested by Maturana. From this perspective, perception is best understood as making contact with the world and exploring the opportunities that environment has to offer—what Maturana calls the "affordances." The affordances are real and perceivable. This is an ecological theory that offers a practical approach to studying cognition across all flora and fauna.

The divide between plants and animals may not be as precise as we thought. We now know that plants exhibit behaviors previously thought to be the exclusive domain of animals. When we think of cognition, we think of behaviors that imply some action, as well as cognitive abilities and a sense of agency. Now, with high-speed cameras, we have been able to look at plants differently and see behavior that we didn't think existed. In the holistic view that Gagliano suggests, the environment itself invites actions and makes behaviors possible, rather than causing them. It does so by providing a continuous flow of information. Perception, then, is a type of self-organizing behavior, and within this context, all living organisms become agents, endowed with autonomy, rather than objects in the mechanically conceived world.

We know there are many aspects of a forest network: the trees, their fungi supporters, insects, mammals, birds, and other animals. They are all involved in an interactive community. And not everybody is a friend or helper. Plants have to defend themselves from parasites, herbivores, and pathogens. They have to be able to distinguish between what's friendly and what's hostile. They can't run away from danger like we can, so they have evolved elaborate defense strategies. They have limited weaponry. They often have to negotiate and try to work out a fair deal. That's part of the spirit of cooperation they demonstrate.

Trees actually have many creative ways of defending themselves. In addition to scent language, consider the following: they grow wax coatings, they have stingers, they release a variety of toxins, they play dead, they call for help from predators like wasps. They are by no means passive.

As science is discovering, trees bond so deeply and so intimately that it's hard to view them as individuals. It may make more sense to view them as a network or a community. They share all resources and give all they can to take care of their forest community. Much of what we have discovered about trees also holds true in grassland communities. Fungi that are associated with grassland plants will also send warnings to adjacent plants when they are being eaten by insects or attacked by pathogens, and they will send food or essential nutrients to neighbors of both the same and different species. When plants are trimmed, those trimmed plants will often send food to healthy neighbors, something that plants also do when they're dying.

This behavior has been observed in many tree communities, where struggling trees will help neighbors even of a different species, particularly when they're dying. Is this altruism? Compassion? How is this orchestrated? Can you have a form of intelligence without a brain and spinal cord? Can you have compassion without a heart?

There have been massive die-offs of Douglas firs in the West in recent years due to drought and insect infestations. What has been observed is that the Douglas firs, when they're dying, are feeding many trees of other species through their root systems with the help of the fungi web. And not just in small amounts. They are dumping all of their reserves into the larger community—resources that they might otherwise have used for reproduction and for their defense. They are stressed and dying, but they are helping their neighbors eat and build defensive enzymes. Why would firs be helping ponderosa pines survive? Do dying Douglas firs intentionally and altruistically give food and defensive signals to ponderosa pines? Is this behavior dictated by the fungi? Is there a pragmatic reason why Douglas firs transfer all the food they can to their neighbors, the resource-hungry ponderosa pines? These are trees they compete with for sunlight. It is one thing to help one's offspring and trees one has befriended and intertwined with for hundreds of years—but these trees normally compete. It would be hard to conclude that there's not some degree of altruism here.[21]

It has been observed that plants are much more sensitive and sensing of the world around them than animals are. As Italian botanist and plant neurobiologist Stefano Mancuso explains, every plant root tip has a tiny region that functions as the locus of electrical signals—the same signals found in human neurons. That means that every root can detect and monitor concurrently and continuously at least fifteen different chemical and physical parameters. He explains why it would not make sense for plants to have the kind of brain we have. The plant would be too vulnerable, so it apparently has a decentralized intelligence scattered throughout the root system. That is why a plant can continue to live even when ninety percent of its roots are cut. According to Mancuso, this allows plants to engage in neuron-like activity, including making mathematical computations and being kind to relatives and friends.[22]

Maybe that also answers the question of why trees in a forest community would keep alive the roots of an ancient mother tree that died some 400–500 years ago. Remember, each tree's roots are connected to the roots of other trees in the forest network. They interact and they communicate. Maybe those elder roots of an old mother tree are part of the forest brain—part of the memory—the heart-brain that holds the collective wisdom of the forest. Perhaps they are like the matriarchal elephant that holds all the collective wisdom of the herd—that knows where to look for food and water when resources are scarce and knows how to keep the community alive in difficult and changing circumstances.

Are trees capable of self-recognition? Do they have a sense of self? It is well known that black walnut roots emit a compound that limits the growth of other roots but not their own. That is why gardeners find it is nearly impossible to grow anything under a black walnut tree. Certain desert plants were found to inhibit the growth of other plants that they came into direct contact with. But they did not impede the growth of roots of other plants of the same species. Does this suggest they could distinguish self from other? Other studies show that plants have recognition systems that ensure they will only mate with genetically different plants.[23]

UK scientist Professor Alison Smith reported to BBC News that plants may use complex mathematics to regulate their metabolism. Mathematical computations are a method plants may use to determine how much photosynthesis they need to do daily, how much sugar they must produce, and for what time period. This often appears to be a more precise computation than the seasonal decisions, like when to drop their leaves. It is a decision that could, if not precisely calculated, result in immediate death.

Sentience is generally understood as the capacity to feel, perceive, and experience. In Western philosophy, it is the ability to experience sensations. Proponents of animal rights and welfare suggest that sentience implies the ability to experience pleasure and pain, and they insist that sentient beings, at a minimum, should not experience unnecessary suffering.

In Eastern philosophy, sentience is a metaphysical quality of all things that requires respect and care, because sentient beings have the ability to suffer and are thus entitled to certain rights. But according to many Eastern religious systems, and among animal rights advocates, trees and plants are not recognized as sentient. Maybe, if today's technology had been available in Buddha's time, more modern Buddhists would reconsider this matter and hold that all life, including our green allies, is sentient, as indigenous cultures have long advocated.

Peter Wohlleben certainly believes that trees are sentient. He would suggest that it is all in the roots. That is, he believes that the roots are where a tree stores experience and that the roots act as the brain of a tree. We know the root system is in charge of all the chemical activity of the tree. In addition to chemical activity, there is a lot of electrical activity in the roots. He says that scientists at the Institute of Cellular and Molecular Biology at the University of Bonn are of the opinion that brain-like structures can be found at root tips. There, they found both signaling pathways and neural-like systems as well as molecules similar to those found in any animal's neural systems.[24]

He chastises us to stop thinking of trees as merely commodities, as objects for our use. He reminds us that we've gotten better at humane treatment of animals and we also need to spare trees unnecessary suffering. He says it's okay to use wood as long as trees are allowed to live in a way that is appropriate to their species, and he stresses that they should be able to fulfill their social needs, grow in a true forest environment in undisturbed ground, and have the opportunity to pass knowledge to the next generation. Some trees in forest communities, he states, should be allowed to grow old and die natural deaths.[25] That does not seem too much to ask.

So, it appears that trees have brain-like structures in their roots that serve neural-like functions. They do not have a brain like ours. They are other. They appear to have complex communication and some type of mathematical functioning. They learn, they remember, and they call on past knowledge to make decisions. They actively acquire and process information from their environment, they have a rudimentary sense of self, and they perceive their world through a variety of senses, both similar and dissimilar to ours.

In addition, they are a part of complex interactive networks, which means that all of their cognitive-like functions are not merely individual; they are communal. We can barely grasp what that means, much less what it would be like to be so intimately—so vitally—connected to all beings in an entire forest community.

By far one of the most fascinating things we have learned about trees is that they appear to act selflessly. They seem to exhibit what we, in our anthropomorphic lens, call altruism and compassion. We humans aspire to this level of loving kindness. In trees, this behavior seems to extend well beyond their kinship networks. These altruistic values do not have an obvious practical value. At least, they don't in terms of how we as individuals define and determine what is practical.

As humans, we tend to think of compassion and altruism as a deep caring for others beyond the immediate self, beyond immediate family. In us, it is a heart-centered aspiration. It is love—beyond the love of self, which many feel is a deep, heartfelt emotion—love that extends to all others. How can you have love without a heart? How can you have intelligence without a brain? What generates the love and caring of offspring in a tree? Why do they take care of their tree friends? Why do trees care for elders? Why would they preserve the roots of a long-dead mother tree?

We know our heart is more than a mechanical pump. It also may be the seat of emotion and feeling. We also know that our mind and heart interact. Our highest functioning—our highest altruism and attributes, like loving kindness—may come from the unification of the heart and mind: the heart-mind. Is it possible that trees and forest networks have a type of heart-mind? Not like ours, certainly, but their own versions. They exhibit behaviors that show a type of intelligence. They exhibit behaviors that show deep caring and compassion. Is there some sense of oneness that is also sent from root to root through the entire wood-wide web? Do they, in some way not definable to our species, intrinsically know they are all part of one organism, one network, one system? It appears they have some type of consciousness, albeit a consciousness quite different from ours. Is it possible that they perceive holistically—that holistic perception is a primary modality for trees within a forest network? I understand that we cannot grasp what that would be like. We cannot grasp it because we are limited by our own conceptual frameworks. We only aspire to occasionally perceive and interact holistically. But we can imagine. For one moment, we can imagine that, pulsing through our veins along with vital oxygen and nutrients, all of which is given to us by our plant friends and that feed every cell in our bodies, is a more ephemeral nutrient that provides us with a sense of oneness—that implicitly connects us to all of life and aligns us with the entire planetary network.

2 FINDING OUR PLACE IN NATURE

> *Ginkgo* (Ginkgo biloba)
> *The ginkgo is a sacred tree in eastern Asia, where it is often planted near Taoist and Buddhist temples. . . . In the wake of the atom bomb dropped in 1945 on Hiroshima in Japan, every living thing around the epicentre of the blast was destroyed. An exception was provided by four remarkable ginkgo trees that survived, and which by the following spring even started to blossom again. . . . Ever since, in Japan, the ginkgo has been regarded as the "bearer of hope."*
>
> *The Meaning of Trees*

We now know that trees have a number of attributes that distinguish them from humans and most animals. First of all, a forest is a network—a community. That association is far stronger than any sense of individuality. A tree works as a part of a whole system, not as an individual among a group of other individual trees. Trees are, therefore, by their nature, cooperative. They seem to share resources with those who are struggling and seem to demonstrate many characteristics that humans aspire toward. They demonstrate what we think of as kindness, friendship, and nurturing. This indicates that they seem to implicitly understand that the whole is greater than the sum of its parts.

While we may wish we were more like trees, it is abundantly clear that we human animals evolved differently and came to very different conclusions. But, you might ask, didn't all life originate from the same source? The fossil record does tell us that plants and animals evolved from a common ancestor. The question is, what happened to us that made us so different? So individualistic and so dominant? How did we as a species, in spite of our many positive attributes, come to harbor the destructive tendencies that have brought us to the brink of the sixth extinction? Why aren't we more like trees? Why don't we think and function as a network? Why did we evolve the way we did? It is a long and complicated story.

Both plants and animals evolved from a single-celled eukaryote. A eukaryote is an organism whose genetic material or DNA is organized into a membrane-bound nucleus. In contrast, bacteria do not have nuclei or other complex cell structures. We believe that eukaryotes evolved when anaerobic prokaryotes, which do not use oxygen, engulfed a type of anaerobic prokaryote that used oxygen to produce energy. For some unknown reason, instead of being eaten, in this instance, the captured organism formed a symbiotic relationship with its host. They made a deal. We believe this is how the mitochondria that power our cells evolved. We believe this is also how chloroplasts, which plants need in order to photosynthesize, evolved—and ultimately how plants evolved. They were engulfed by a eukaryote who also engulfed a chloroplast. It is believed that the eukaryotes that did not have chloroplasts became animals and fungi. Some, of course, remained single-celled organisms. But those that had chloroplasts became plants. The term endosymbiosis means that one cell ingests another and fails to digest it. That is how we got here. This may be the earliest form of cooperation.

While we came from common ancestors, plants and animals did not evolve from each other. Animals are actually more closely related to fungi than to plants. Both plants and animals are made up of multi-celled eukaryotes. Their cells contain nuclei and organelles, which are nothing more than specialized cells—organized structures living within a cell.

Based on fossil evidence of simple single-celled and more complex multicellular organisms, we believe that the first eukaryotes evolved approximately 2.7 billion years ago, at the end of the Precambrian Era. We call this era the Protozoic.

Both plants and animals evolved in the oceans where they lived together for at least 600 million years before moving on to the land. At this time in our early history, the land was too toxic for life as we know it. It did not have a protective ozone layer, so it was inundated with deadly levels of radiation. The ozone layer formed once photosynthesis developed and raised the oxygen levels. Only then could early life venture out of the sea.

Most plants photosynthesize, but they are not the only organisms that do so. Other eukaryotes like algae photosynthesize, as do several groups of bacteria. We believe that plants evolved from a type of freshwater plant-like algae that also photosynthesizes.

Plants were the first to venture out of the ocean. They probably initially tried to anchor themselves near the shore. The Earth then resembled the rocky surface of Mars; it was not lush, soft, and green. These early plants gradually transformed the land base. But it was not an easy life. There was little for them to root into. They had to deal with new forces: gravity, wind, waves, tides, and a fierce sun. They had to adapt. Over time, they developed vascular structures that gave them woody support systems.

There are two main groups of plants: vascular plants and nonvascular plants. Nonvascular plants are things like liverworts and mosses, which do not have stems that conduct water. All other plants have specialized vascular tissues, phloem and xylem, which help them transport minerals, sugar, and water through their bodies. The oldest vascular plants appeared in the middle Silurian Era, about 409 to 439 million years ago. The oldest nonvascular plants appeared much earlier, in the Ordovician Era. Of course, this is based on the fossil record, and non-vascular plants simply don't fossilize well so they may have appeared far earlier. Some 290 million years ago, large, complex seedless plants were dominant. Many were fern-like, grew to the height of trees, were up to ten feet wide, and covered much of the globe. You can picture them like giant, crunchy celery stalks. They grew in swampy areas, and when they died, they sunk below the surface. Gradually, their carboniferous plant matter was compressed and transformed to peat and coal. So in a way, we can blame our climate crisis on them.

These plants were replaced by gymnosperms. Gymnosperms are seed-bearing plants, including conifers, ginkgo, firs, and woody shrubs. They dominated for the next 200 million years, and in turn were replaced by flowering plants, angiosperms, in the middle of the Cretaceous Era. Flowering plants, many of which bore fruit, were the most widespread plants on Earth for at least a hundred million years. From their emergence to the present, plants have existed as a part of extensive communities. They did not venture off as individuals. They evolved from endosymbiotic eukaryotes, as did animals, but they developed into cooperative networks and forest communities in which cooperation and sharing were the norm.

These early plants made it possible for animals to venture from the ocean. At every stage in the terrestrial land base, plants came first and created new ecological systems that allowed animals to get a foothold in a new world. Once plants settled into the swampy areas, insects were the first to join them on the land base. Some got to be huge. The fossil records indicate that early dragonflies had a wingspan of eighteen inches and millipedes could be over eight feet long. They were followed by early amphibians who preyed on these guileless creatures. Some of these amphibians were twenty feet long. Major evolutionary leaps in plants were often followed by leaps in the animal kingdom. When plants adapted and colonized new areas, animals adapted to these new ecosystems and thrived. As plants ventured into more arid valleys and mountains, amphibians could not easily follow, but reptiles adapted and soon became the predominant land species. When flowering plants (angiosperms) evolved, so did pollinator insects. Their mutual success depended on a complex symbiosis, called by some "the survival of the most cooperative." Then came birds and mammals, who fed on their seeds and flowers, and they too thrived. The plant world was less affected by the turbulence of Earth changes that repeatedly led to mass animal extinctions. They continued to adapt and diversify, and as they did, new life followed, generally insects first, then other predators like frogs and songbirds that evolved to feed on the insects. When the climate got dryer, the forests became smaller, and grasslands and savannahs evolved. These grass-based plant communities made a new niche for mammals like rodents, who in turn supported snakes. Then came grazing animals. Plants led the way creating new ecosystems for animals at each step on our long path of evolution.[1]

Although both plants and animals evolved by natural selection, adapting to make some species better suited for their environments, animals had a completely different path—a seemingly tougher path than plants. They did not photosynthesize. They could not produce their own food. There was no manna from heaven. They relied entirely on plants, their fruits and seeds, and the animals that ate plants. They had no choice. You either ate plants and other animals or were eaten by others. It was the law of the jungle. You were predator or prey. Your survival demanded strategies far different than those of plants.

Unlike plants, animals are mobile. They needed to be mobile to search for food sources and to avoid becoming someone else's dinner. So, they quickly developed behaviors that required highly aggressive attitudes. They needed to learn how to defend themselves. While they congregated in groups to avoid becoming prey, they also had to evolve more individualistic behaviors to survive.

We believe that multicellular animals evolved about one billion years ago. Rocks in China that date back 600 million years have been found that show fossil embryos. Animals also evolved in the sea and remained there for at least 600 million years, along with plants. We believe there was great variety among animal forms in early life. The earliest fossil record also indicates that there is considerable evidence of animals that showed predatory behaviors quite early. We therefore believe that complex prey-predator relationships existed very early in our evolution.

There was an explosion of life in the Cambrian Era that resulted in many diverse animal forms. We believe this happened because there was more oxygen in the atmosphere about 200 million years ago. The existence of more oxygen allowed for higher metabolic rates, so larger, more complex body structures evolved. Over time, animals were able to develop hard skeletons made of calcium carbonate. 500 million years ago, animals were mostly invertebrates. The first bony fish and sharks evolved a little later, over 400 million years ago. By then, both plants and animals were colonizing the land masses.

In the early seas, predatory strategies were probably used by both eukaryotes and bacteria. This included swimming up to a smaller cell and engulfing it, forcing its way into larger cells and digesting them from the inside, and forcefully digesting them from the outside. It was a grab or get grabbed lifestyle.

We learned to defend ourselves very early on. 740 million years ago, tiny amoebae developed tough skeletons.[2] Hard-shelled fossils at least 550 million years old have been found that show scars and marks of predation.

From the fossil record, we know that the first predators evolved about 500 million years ago. They had two essentials that set them apart from other early life: jaws and teeth. Recently, scientists at the University of Erlangen-Nuremberg used an electron microscope and an x-ray spectrometer, which can measure the composition of materials, to study fossils of a conodont—an eel-like creature from the swamp of early life that not only had teeth but could repair and regrow them.[3] Teeth were very important, as were grabbers: huge limb-like structures used to grab and stuff prey into a predator's mouth.[4] Brain size did not matter much. Early "mega-predators" such as anomalocarids evolved giant mouths to suck up worms. It did not take smarts—just good avoidance skills, teeth, and grabbers.[5]

Among the earliest terrestrial vertebrates were arthropods: invertebrates with an exterior skeleton, similar to present-day lobsters, shrimp, spiders, centipedes, and wingless insects. In these early days, some 500 million years ago, the largest arthropods were over a meter long. That was huge for the time period. They were the killer sharks of their age, with radiating teeth and huge grasping appendages. Smaller animals did not have a chance.

Some early arthropods also had a type of eye. Apparently, both arthropods and hydras (a genus of freshwater organisms like jellyfish) developed rudimentary eyes about 600–700 million years ago. By eyes, we mean a very simple type of selective signal enhancement tool that was so primitive and basic it did not require any type of central brain. This eye was no more than a system of touch sensors that helped this very primitive organism sense where to direct its giant claw-like appendages. Nevertheless, this was a scary predator, and single-celled and early multi-celled life forms swimming in the primordial soup did not want to get on its radar.

Consciousness in animals developed slowly. Over the last ten years or so, a new theory has arisen from evolutionary biology called the Attention Schema Theory. This theory suggests that consciousness developed from the simple problem of information and sensory overload.[6] According to this theory, developed by neurobiologist Michael Graziano, there was so much information to process that a rudimentary nervous system became necessary. In vertebrates, this "brain," like the arthropod eye, allowed for the processing of some signals at the expense of others. If this theory is correct, consciousness evolved gradually over that last 500 million years, beginning in early vertebrates. This simple eye allowed a few signals, what we now call selective signal enhancement, to rise above the noise.

Arthropods were followed by amphibians and then reptiles. We believe that mammals evolved from a lineage of reptiles some 248 to 286 million years ago. At some point, early animals developed a centralized controller for attention. This was a jump forward, as animals could then coordinate among all their senses. This is known as the tectum, and it is found at the top of the brain of primitive animals. According to Graziano, it acts like a satellite dish, gathering signals from the senses and processing them to determine what is important, like: "Can I eat it? Will it eat me? Is it my girlfriend?" The tectum is found in all fish, birds, reptiles, and mammals but not in invertebrates, and it is used to predict and plan to some degree. It is still the largest part of the brain in fish and amphibians.

Then, Graziano tells us, about 300 to 350 million years ago, reptiles evolved and developed a new brain structure called the wulst. Mammals and birds inherited it from the reptiles we evolved from, but in mammals it is called the cerebral cortex. Scientists refer to it as a spotlight. It can shift your attention to sounds outside or behind you, to thoughts and memories, to processing information. This may have been the beginning of abstract thought. Scientists also believe this may have been the beginning of consciousness as we think of it. As Graziano says, "Somewhere deep in the brain is something primitive that is computing a semi-magical self-deception. This mysterious process allows something as primitive as a crocodile to grasp ahold of something intangible, to understand it and to remember and to respond." From an evolutionary biology perspective, this is a brief review of how scientists now believe abstract thought and consciousness may have developed.

We know our brain evolved from our reptilian and mammalian ancestors based on the survival dictates of the predator-prey world they lived in. The next critical part of the brain to evolve in mammals was the limbic system. While the reptilian brain controls basic vital functions and compulsive behaviors, the limbic system records memories and agreeable or disagreeable sensations, and it is responsible for emotions, value judgments, and unconscious behaviors. The limbic system contains the amygdala, which appears to be the source of aggression and fear. Another part of the limbic system, the hypothalamus, regulates responses to pain and pleasure, sexual satisfaction, and anger and aggression. It also regulates the autonomic system. One part of the autonomic system is the sympathetic nervous system, which controls blood pressure, heart rate, sweating, and breathing. This is where the fight-or-flight response comes from. This essentially automatic response is associated with running from danger and preparing for violent behavior.

So, when our ancestors saw a saber-toothed tiger—which incidentally had teeth ten inches long—a huge snake, or a stranger, this is what would happen: the sympathetic nervous system would dilate their pupils, stimulate their sweat glands, dilate blood vessels in their large muscles and constrict other blood vessels, increase their heart rate, open up their bronchial tubes, and inhibit secretions in their digestive system in order to maximize their body's ability to fight or flee. This also happens now far too frequently, with uncontrolled road rage and other aggressive, act-first, think-later responses.

If this reaction continues, the adrenals release adrenaline, which causes similar effects that stay in the system longer. The parasympathetic system, rooted in the brain stem, eventually brings the body back from a state of emergency. But once adrenaline is released, things can become disastrous, even deadly, very quickly. These hormones were critically important for our early survival. We needed fast reactions. As we will see, trees and all higher plants produce similar compounds that are thought to regulate their behavior. They do not have a limbic system or cerebral cortex. They still respond to threats, but in different manners. They can't run away or fight back as we do. They defend themselves but also may have learned to be creatively cooperative.

Modern mammal groups didn't appear until about fifty-five million years ago, the same time as the oldest-known primate fossils. We believe that primates evolved from a nocturnal, insect-eating, rodent-like mammal similar to a shrew. Hominids enter the fossil record in the Pliocene Era, some 1.8 to 5.3 million years ago.

Homo sapiens and our close relatives have another brain: the cerebral cortex. This brain helped humans, in whom it is the most advanced, develop more complex skills, create tools, and live together in social groups.[7] It allows for fine motor movements, is responsible for our hearing, visual, and language skills, and processes all higher functions, like learning, thinking, planning, judging, and moral reflecting.[8]

Hominids diverged from other primates about five to seven million years ago. The earliest humans appeared about 400,000 years ago, and the earliest known modern humans date back to 170,000 years ago. We know that our earliest recognizable modern human ancestors lived in Africa about 160,000 years ago. They later migrated to Europe and Asia and displaced other hominid species that were living there. The appearance of complex cultural behaviors like jewelry-making and cave-painting didn't occur widely until 30,000 to 40,000 years ago. Some commentators believe that this followed the development of a highly sophisticated language.

Animals similar to humans have been around for about two and a half million years. There was nothing to indicate the species *Homo sapiens* might evolve into something unusual. For a long time, we had relatives, including *Homo floresiensis, Homo denisovan,* and Neanderthals. We all had large brains and walked on two legs. Our hands were able to perform intricate tasks. Around 300,000 years ago, early human-like species first used fire.

Neanderthals lived in relative harmony with their environment for 430,000 years. Floresiensis were around for about 190,000 years and did not manage to destroy their environment. Likewise, Denisovans, who lived in the high Tibetan plains, coexisted with other life forms for about 160,000 years.

There is evidence that these earlier human variants were killed off by their more aggressive sapiens cousins. Others say that sapiens simply out-competed Neanderthals and the other human variants. It seems that wherever they went, sapiens left a trail of extinction. They were savvy enough to figure out how to migrate to Australia some 45,000–60,000 years ago. There, they were immediately on the top of the food chain. Numerous large marsupial creatures had evolved in isolation in Australia. Within a few thousand years of the arrival of humans, twenty-three of twenty-four of these species vanished as a result of humans. Humans repeated this pattern of destroying all large fauna as they settled in other parts of the world. When they entered Siberia, the large fauna became extinct. When they settled in America some 14,000 years ago, large mammals, like the saber-tooth tiger and mammoth, soon disappeared.[9]

We don't know why our species, *sapiens,* became a dominator species. We do not know what made us so different. But, between 30,000 and 70,000 years ago, something happened. Many researchers believe that there was a revolution in *sapiens'* cognitive capacity. Early *sapiens* essentially became like modern humans, with the ability to be creative, to think symbolically, and to dominate. There is considerable debate about what caused this rapid transformation.

Richard G. Klein, a Stanford archaeologist, believes there was a neurological change that occurred in *sapiens'* brain. He says that about 50,000 years ago a random genetic mutation occurred that rewired our brains. He theorizes that this may have allowed for advances in how early humans were able to speak and perceive.[10] It also altered how we viewed nature and social circumstances, giving us the ability to manipulate both nature and culture.

Others believe that this change occurred much more gradually and that it was not the result of any kind of genetic mutation. We know now that Neanderthals also made jewelry, complicated tools, and structures. Those who don't buy Klein's theory say it's just as likely that human evolution is not the result of some genetically altered shift but rather due to sharing of information, mating, and exchange of cultural behaviors between multiple populations over time. We know humans interbred with Neanderthals and our other cousins. Neanderthals had bigger brains that were wired differently than *sapiens*. They were believed to have far better visual, spatial, and movement skills.[11] There is no evidence these extensive innovations in thought, action, and behavior could have been induced by ingesting any known type of psychotropic, or other mind-altering substance, as the effects were permanent and were transferred to offspring. We simply do not know what other intervention, if any, might have taken place that led to the higher, more complex brain functions we now have.

Whatever occurred was too rapid for ordinary Darwinian descent with modification—that is, simply passing traits from parents to offspring. It may be unsettling, but we don't really know what happened. Nevertheless, a huge and revolutionary change occurred that prompted *sapiens* to rather suddenly transition to modern cognition and behavior. The development of a more advanced sense of language was undoubtedly part of it. But we simply do not know what part new language skills played in this transition or how we suddenly leaped forward.

Some unknown circumstance led to a rather sudden jacking up of *sapiens'* brain. Perhaps we hatched multiple new neural pathways. Perhaps some sort of rewiring and magnification between existing neural structures occurred. Whatever the cause, the result made us extra savvy and the dominator species. It also seemed to spark a creative fire, allowed us to plan ahead, and helped us develop cultural traditions like storytelling and extensive social behaviors.

Up until that point, the development of our brain and neural system followed a slow and linear evolutionary process. It evolved incrementally based on the necessities of survival. Humans were not at the top of the food chain. They were a vulnerable species that probably served as prey far more than as major predator. They had limited shelter from the elements and the very real dangers of their world. They were constantly in motion in search of food. They were constantly on guard. The hardwired instincts they had developed over a long evolution from reptile to mammal to human triggered rapid-fire bursts of aggression, domination, and violence: behaviors that seemed to be essential for their survival.

Our brain has not changed very much. Normally, a species has thousands, even millions, of years to adapt to new circumstances. Everything happened very fast with *sapiens*. And there is no doubt that our hardwired survival brain has not had time to catch up with the reality of life in the twenty-first century.

The recent popular book *Sapiens: A Brief History of Humankind* by historian Yuval Noah Harari discusses the cognitive revolution that occurred some 50,000 years ago. Harari emphasizes that we developed the ability to think abstractly; that is, *sapiens* developed the ability to believe in things that existed solely in the imagination. He cites Klein's theory of a random genetic mutation as the cause. This, he says, led to the beginning of all myths, including those about gods and empires that we have developed and continue to live by. He believes that once humans developed the ability to believe in stories and myths collectively, they were able to live together peacefully in larger numbers. This, he says, led over time to the development of large-scale human quasi-cooperation systems, including religions, political structures, trade networks, and even legal institutions. He traces them all to the emergence of the new human cognitive abilities that allowed for the development of abstract thinking.[12]

As a result, humans developed beliefs and mythologies that placed them outside of nature and that anointed them as the dominant species. We don't know exactly when or how this happened, but it's pretty clear that at some point, *Homo sapiens*, of their own volition and on their own two feet, walked out of the Garden of Eden.

Harari says that most scholars agree that early humans are believed to have been animistic, just as most ancient foragers were. Animism is considered the world's oldest religion and is still an almost universal belief in contemporary hunting and gathering societies.[13] Animism is a belief that places animals, plants, and phenomena on equal footing. All beings are part of a whole. Animists believed that all species, plants, and animals have awareness, feeling, and can communicate. Animistic thinking would tell us that we are not separate from our environment and that we are not separate from other species. We are all connected and a part of an interconnected whole. From this perspective, other species are not simply here to provide for our needs. Nor did existence evolve solely around our species.

Animism is not presently considered a world religion. But aspects of animism survive in many belief systems, religions, and traditions. Animism preceded the development of theocratic religions that place humans outside of nature and insist that humans are dominant by divine right.

Harari says that *sapiens* developed the ability to perceive and talk about things that they had never touched or smelled. They developed the ability to speak in fictions—an ability that's unique to our human language. He says *sapiens* could not only imagine things, but they could imagine things collectively.[14] He says our shared mythology allowed *sapiens* to work more cooperatively in large numbers and eventually to create and live in complex megacities. Ultimately, these new stories provided the fuel necessary for us to colonize, conquer, and manipulate our world.

And this ability to talk about imagined things also led to the development of hierarchical relationships between humans and fictional entities called gods. As humans developed the ability to live together in crowded cities and empires, it was these invented stories about gods and motherlands that held them together and provided them with a functional social order.[15] Harari emphasizes that these cooperative networks were imagined orders, with social norms that were based on beliefs, not reality. They were seldom voluntary or egalitarian. They did not bring happiness, nor did they bring well-being. And because they were not extrinsically real, huge effort was made to sustain laws, customs, and procedures to maintain these imagined orders.

He defines religion as a system of human norms and values founded on the belief of a superhuman order. In contrast, when people were more animistic, they shared beliefs and norms that took other beings into consideration. During the agricultural revolution, which began about 9500–8500 BCE, farmers learned to manipulate plants and animals. It also meant that we, not plants, were now modifying and changing ecosystems. These behaviors revolutionized how we thought of other species. This demystification of the natural world accelerated some 10,000 years ago. Other species, both plants and animals, were no longer equal members of a spiritual group; they became property.[16] The role of the new gods was to mediate between humans, plants, and other animals. A new mythology required a new bargain. Through devotion and sacrifice to these new gods, humans were granted mastery over plants and animals. The older animistic practices were gradually lost, except in traditional and indigenous cultures.

The new early religions, which were patriarchal—that is, cut off from the feminine, holistic source—led to the monotheistic myths that were encoded into our culture and helped engender our present environmentally destructive beliefs, institutions, and policies.

A classic example is found in the Christian book of Genesis, which tells us that nature is subservient to man and directs us to subdue the

Earth and have dominion over "the fish in the sea, the birds in the air, over the cattle, and over all the wild animals of the earth, and over everything that creeps on earth." Many still believe this language is a divine edict that gives humans dominion over the entire Earth.

American fundamentalists and Pentecostals are focused on stories of the afterlife, and many believe environmental concerns are a distraction. Some who hold apocalyptic beliefs welcome the idea that the environmental crisis might lead to the destruction of the world as we know it.

As we have seen, one set of myths was replaced by an even more dissociated set of myths. Sadly, these misconceptions are still being used as a justification for massive environmental destruction.

What remained of the vestiges of an interconnected sense of reality that these new theocracies didn't destroy was toppled, perhaps inadvertently, by the emergence of scientific thinking. The Age of Enlightenment, which led to modern empirical and scientific thought, built up over time and was gradually accepted by many as the only logical path to truth. But it had a dark side. The study of nature became divorced from human beings. In the seventeenth century, Rene Descartes proposed that the entire world was to be reduced to what we could "measure and number." Francis Bacon stated that humans were distinct from the natural world, therefore implying we could essentially abuse the natural world with impunity. These concepts became the basis of a new way of viewing the world and our place in it.

That is not to say that science is just based on another set of myths. Science is a way of perceiving and acquiring information and coming to conclusions. The scientific method involves careful observation. It begins with a hypothesis. This hypothesis is tested, refined, or eliminated based on the experimental findings and verification.

Science is not the problem. Science, however, does rely on the concept of reductionism. Reductionism is an approach to understanding complex things by breaking them down into simpler fundamental things, and it leads to the idea that the whole is nothing more than the sum of its parts. This concept underlies much of modern science, including physics, chemistry, and cell biology. Of course, the wonderful new research on forest networks relied on reductionism. There is nothing wrong with the approach but it should not be overused or seen as the only way to view the world.

Reductionism reinforces other problematic perceptual viewpoints, such as dualistic thinking, them vs. us, good vs. evil, black vs. white, man vs. nature, etc. Reductionism and dualistic thinking both contribute to another false perception: the subject–object dichotomy. This abstraction tells us that the world consists of objects that are observed by subjects, who have unique consciousness and internal perceptions. This, of course, means those objects, being all of nature, are separate from the subjective self and are merely "objects," which can be commoditized for the subject's use and enrichment.

These ingrained concepts are methods used by us to abstract and minimize the wonder of the natural world and bring it under our dominion. In reality, we are discovering that the whole is greater than the sum of its parts. There is no real them vs. us. There is no intrinsic justification for objectifying and commoditizing the complex world we are a part of. The relationship between the whole and parts cannot be disentangled. We are beginning to understand that the essential nature of life on our fragile planet is holistic and inseparable.

Our ability to think abstractly allowed us to internalize limited perspectives and biases, which provided us with justification for the subjugation of nature. When we consider the emergent properties of complex systems, these perspectives fail. The good news is that science is not just reductionist; it is also open to looking at things in a more holistic manner. Increasingly, scientists are finding that reductionism alone is of limited use in studying organizations of greater complexity, including living cells, our brain and neural networks, ecosystems, and societies.

Science can be faulted for supplying the old story that says nature is like a machine that can be broken down into its parts. It is mechanical. It is a complex machine, but it is just a machine. It is not alive. It is not conscious. And it is separate from us. It is just an object or commodity intended for our use. That story no longer holds up.

Now science is suggesting a new story. The Gaia hypothesis, formulated by chemist James Lovelock and codeveloped with microbiologist Lynn Margulis, proposes that living organisms interact with their inorganic surroundings on Earth to form a synergetic and self-regulating complex system that maintains conditions for life on our world. The Gaia hypothesis helps us understand the biosphere, the evolution of organisms, the stability of global temperatures, the salinity of seawater, atmospheric oxygen levels, and other environmental variables that affect the habitability of the Earth. This hypothesis suggests that organisms coevolve with their environments.[17]

Some versions of the Gaia hypothesis have been used to create a new mythology: a new story that tells us that all life forms are considered to be part of one single living planetary being called Gaia. From this viewpoint, the entire Earth—its atmosphere, its seas, and its terrestrial crust—are the results of interventions carried out by Gaia through the coevolving diversity of living organisms. This new story is actually a very old story. Gaia is the Greek version of Mother Nature, of Mother Earth. It is also the basis of the older, more animistic beliefs of our predecessors.

Scientists who are now studying the complexity of the forest networks are teaching us that nature is composed of interdependent systems within systems. Their research suggests that an inherent intelligence may exists in plant networks—an intelligence that we didn't know about.

More holistic approaches to science involve research that emphasizes the study of complex systems. These systems are approached as coherent, complex wholes. Component parts are best understood in context and relationship to one another and to the whole. This way of approaching science focuses on the observation of a specimen within the ecosystem first before breaking it down to study the parts of the specimen. It is also based on the idea that the scientist is not just a passive observer of an external universe but rather a participant in the system. Holistic science is best suited for subjects like ecology, biology, physics, and the social sciences, where complicated nonlinear interactions are the norm.

There are many ways in which this holistic approach to science is making its way in our world. Permaculture and sustainable practices are integrating system-level approaches to agriculture and land management. Organic farming is a holistic approach. A holistic approach was applied by quantum physicist David Bohm in his Theory of Implicate Order. It is applied to medicine as holistic or alternative approaches. An evolutionary biologist, Elizabeth Sahtouris, has written extensively on the need to view biology more holistically.

The so-called cognitive revolution that led to modern cognition fueled the development of the rational mind, subject–object perception, and dualistic thinking, as well as a sense of separate ego-based self, all of which are useful human attributes. But these same attributes, used without the balance of a deeper understanding of the complex systems in which we act and interact, also led to all the resultant modern dis-associations, disenchantments, and delusionary behaviors we humans exhibit. We lost our place in the whole. And it may have been genetic mutation that got us into this mess. If so, we may need to consider whether that mutation may have also been a disastrous error. As a result of this possible flaw, we now have a new geologic era, the Anthropocene, named after us. We humans are so rapidly changing and destroying our environment that we have already caused the extinction of hundreds of thousands of species, many of which had yet to be named. We may, in fact, be on track to cause our own extinction. Should the climate change we have caused continue unchecked and rise to even four degrees Celsius higher than it is at present, we can expect that the world as we know it will rapidly become uninhabitable. This will happen so quickly that the population of *Homo sapiens,* now at 7.7 billion, may soon be less than half a billion, if we survive at all. That will be the ultimate test of the value of the mutation (if it was a mutation) that led to the cognitive revolution. In the long run, did it help us thrive and survive in the complex ecological system we are a part of? We shall see.

Whether it was a rapid mutation or more gradual change in brain function that gave us the cognitive revolution and the ability to think abstractly and in a manner that minimized the rest of nature, from an evolutionary viewpoint, it was still a rapid shift. It is likely this shift may have also engendered the concept of a separate self—the ego-based self. The ego is a mental construct that habituates our inner patterns of thinking, feeling, and acting. We view our inner self as separate from the outside world.

In a world that was based on predator–prey behavior and flight-or-fight instincts, the development of an ego and the defensive behaviors and patterns it requires led to more aggression and violence. But it may have been essential for survival.

A distinction, however, is important. The ego is not the same as self-awareness. A sense of self-awareness develops in children by the age of three. They can identify themselves in a mirror. They can think of themselves from the perspective of another. And they begin to develop a self-concept made of thoughts and feelings about their inner self. Of course, many animals, including primates, elephants, magpies, and dolphins can also identify themselves in a mirror and have self-awareness.

However, it seems pretty clear that most animals have a degree of self-awareness. Even cells have a boundary. To survive, all animals have to know where they stand in relationship to other family members and other species. They need to know who is prey and who is predator. Even plants may have some sense of individual identity, although there is no evidence they have an ego. They seem to know which saplings are theirs, and they can definitely distinguish between their roots and the roots of others.

But only humans have an ego, and only humans have egotism, that is, the practice of thinking about oneself excessively because of unwarranted self-importance. Many humans collectively have developed an inflated sense of separate self that is proving not just selfish but often pathological. It underlies our current climate crisis and much of our polarization. Too many humans think everything evolves around them, their lifestyle, their tribe, their nation, and their social group. Unfortunately, as we see daily, many will take any action necessary to preserve their false view of themselves and their social groups, without regard for others and certainly without regard for the natural world. This extreme degree of anthropocentric egotism is increasingly pathological.

Even those of us who are not classically egotistical or pathological think of the separate self as real. We do not recognize that it is just a construct made of our likes and dislikes and that it functions to make us feel separate from the world. This false sense of separate self, which can spawn extreme self-importance, is built on estrangement of self and nature. It has led to our viewing the natural world, including all trees and plants, as nothing more than the source of commodities that we can compile, store, and hoard to fill our inner hollowness. This is both the cultural obsession and a pathology of our modern world.[18]

One of the biggest obstacles to effective action on climate change is our collective selfishness, our collective egotism. Some research has confirmed that many people do not want to endure any present pain or sacrifice to avert climate change, regardless of the effect it will have on their children and grandchildren, much less the rest of the natural world. Information is not enough; we need to be enticed to change. Jennifer Jacquet of New York University suggests that we consider using shaming strategies that have been found to be effective in game theory to encourage environmental social change and lessen collective and corporate egotism. There has been considerable research suggesting we will need considerable nudging and incentives to make the behavioral changes necessary to adopt more climate-friendly, greener practices. But this research also indicates that once these nudges and incentives are in place, the new behaviors become second nature.[19]

The paradox is that the same characteristics and perceptions that came with the cognitive revolution, and that can be so destructive, have also given humans a singular brilliance. This brilliance has allowed our species to make unprecedented leaps, from the spectacular artwork created in Chauvet and Lascaux some 17,000 to 35,000 years ago to the pieces housed in the Louvre and the Met. In a matter of decades, we went to the moon and then explored the outer regions of the solar system. We have invented the internet and live in an era of immediate global communication. In a very short time, we have dramatically improved health care and education worldwide. Poverty and starvation have dramatically lessened. We have become hyper-mobile and can traverse the entire planet in hours. And we are also, just as rapidly, destroying our planetary home.

This is how we got into this mess. It was a long journey through time. In truth, we are not like trees and plants, at least not much. We did not develop the ability to photosynthesize. There was no free lunch. From the onset of our humble beginnings, we were always on someone else's lunch menu. To survive, we needed clever strategies and aggressive, individualist behaviors. We needed smarts. This long, slow process was not very pretty and not all that orderly. Over a long evolutionary time period, we developed complicated neural structures that heightened our competitive advantages. We modern humans have great skills and much potential, but at our core we are still often fear-driven, with hardwired instincts for aggression, violence, and domination. These instincts are no longer appropriate for the environment we find ourselves in. We cannot unwire our brains. But we can learn to balance and control primal instincts and emotional patterns born of another time. We can learn to be mindful. We can soften and tame the wild beasts we carry within us.

As for the other dysfunctions we host, such as false stories of false gods and a myriad of other myths and methods of perceiving that no longer entirely serve us, it is time to take a hard look. I am not suggesting that we give up our belief systems. I am suggesting that we examine the stories embedded in both conventional religions and the remnants of more ancient teachings. Do they still serve us? Do they deepen our understanding and integration? It may be appropriate to think of some of our outdated beliefs, habits, and perceptual patterns as parasites clinging to their unwitting hosts, sapping our vitality and our goodness, keeping us confused and anxious. It is time to shake them off. It is both appropriate and timely to conclude that many of our old stories no longer serve us and have become deadly. We need new narratives—new ways to view ourselves. That is the easy part. The hard part is the deconstruction of the perceptual blinders we have developed. We have scars that have been imprinted on our collective psyche as a result of unrestrained behavioral patterns and cultural biases, such as reductionism, dualism, rationalism, the subject–object dichotomy, and self-importance. These patterns and biases are not necessarily bad or wrong. They are just overused and need to be softened and balanced. They need to be integrated within a deeper understanding of our actual place within the whole of life.

The good news is that the ways of thinking and perceiving that have shaped many of our destructive past behaviors are only mental constructs. We have been boxed in by these limited perceptions and ideas for far too long. It is time to break free. Deconstruction is hard work. It must happen on the individual, the collective, and, yes, the mythic level. We need to remember who we are, not just where we came from.

With the same singular brilliance that got us into this mess, we can reframe our understanding of our world and our place in it. This is the emergent new story in which our old destructive patterns will no longer find support.

Not that long ago, *Homo sapiens,* so full of himself, seemed to walk right out of the Garden of Eden. Guess what? That is just a story. A better story is that we never left the Garden. We have just been very confused. We got lost in a maze of our own hubris, a maze of our own creation. We have become dis-eased—that is, disassociated and disenchanted. And guess what? There is a cure. It has been there all the time, beckoning us, softly calling. It begins with a walk in the forest.

3
HOW NATURE HEALS US

> *Willow* (Salix)
> *Salicylic acid—the anti-inflammatory and pain-relieving ingredient in aspirin tablets—was discovered in the salicin in willow bark. . . . Since ancient times willows have been associated with the moon and the feminine. . . . Prophecy and divination (particularly with the help of water, as in scrying, or looking into a well, brook or bowl of water when in a trance-like state) as well as healing, white magic, poetry-writing and music-making, activities were the domains of the moon priestesses in the willow groves.*
>
> *The Meaning of Trees*

From the time I was a small child, I always felt refreshed, calm, and joyful in nature. As a toddler, I spent contented hours in our backyard spruce grove. As soon as I was able to go off on my own, my preferred companions were my tree friends in the neighboring woods. I do not remember a bad day with them. We did not fight or argue. No matter how troubled I was when I arrived, I always left balanced, a little exhausted from my adventures in their presence, and somehow healed.

There was nothing specific that happened in my tree encounters.

We coexisted. I was just a child playing in the forest. I think that is the key. We don't have to do anything. We just have to be—be present in nature's presence—and balance and healing happen. As simplistic as this may sound, there is a great deal of science to back it up. And as it turns out, a walk in the forest is more than just a walk.

Two-thirds of the country of Japan is covered with forests. The Japanese have a cultural tradition of spending time in nature. This in part stems from their predominant religions, Shinto and Buddhism. According to Qing Li of the Nippon Medical School, many practitioners of these religions believe that the forest is the realm of the divine. In the ancient Shinto teachings, religious spirits, which are representations of the divine, known as Kami, are found in trees, rocks, the wind, and water features. In other words, everything is alive and animate. Li states that many Zen Buddhists see the whole world as vibrant and spiritually alive.[1]

These beliefs are also embedded in their folklore and mythology. While many cultures have a history of animism and tree worship, it should come as no surprise that Japan is the forerunner in research that explains why people of all cultures feel so much better when they spend time walking in a beautiful forest. In Japan, this tradition is called "forest bathing" or *shinrin-yoku*. Forest bathing is not just hiking or jogging through the woods. It involves opening our five senses, fully taking in the forest atmosphere, and connecting with the natural world. But really, all we have to do is breathe. We don't have to do anything but spend time in the forest. It is that simple.

We know that forests produce oxygen and purify the air. And, believe it or not, trees are natural air filters. Their leaves and needles constantly catch airborne particles. A square mile of forest can catch up to 20,000 tons of harmful material a year. Trees filter out pollutants, pollens, and dust, including acids, toxic hydrocarbons, and harmful nitrogen compounds.[2]

During the summer months when all trees are photosynthesizing, each square mile of forest releases about twenty tons of oxygen in the air. This incredible amount of oxygen is a byproduct of photosynthesis. Humans breathe about two pounds of oxygen a day.[3] So, a walk in any forest on a summer day will bathe you in abundant, clean oxygen. But that is not the primary benefit of forest bathing.

In 1982, when he was the director general of the Agency of Agriculture, Forestry, and Fisheries of Japan, Tomohide Akiyama coined the term shinrin-yoku. It was part of a campaign to get more people to visit forests for health benefits and was also intended to encourage the protection of forests.[4]

In 2004, Japanese researchers began to formally explore the connection between human health and time spent in forests. Li was one of the first to determine that forest bathing can boost the immune system, increase energy, decrease anxiety, depression, and anger, reduce stress, and lead to relaxation.[5] That seems unlikely, but as we shall see, scientists around the world have verified these results.

Could the increase in oxygen have this effect? Certainly. Breathing filtered air rich in oxygen is beneficial, but that is not what is going on. But before I explain why spending time in nature improves our health, I would like to share some of the details of the research results Li and others have consistently found.

Remember the limbic system that is found in all mammals? It is the home of the fight-or-flight response. This is part of the sympathetic system, which makes your heart rate and blood pressure increase. It is the cause of severe stress reactions. Your body dumps cortisol and adrenaline, which are stress hormones, into your system. It is a hardwired sympathetic system response that is balanced by the parasympathetic system, which in turn lowers cortisol and adrenaline levels, suppresses the flight-or-fight system, and lowers blood pressure and heart rate.

Most people live in stress-inducing urban environments. As a result, their bodies produce considerable cortisol and adrenaline. Their sympathetic system is hyped up. The researchers in Japan found that when people spent time walking in forests, their cortisol and adrenaline levels dropped, the flight-or-fight reflex was suppressed, their blood pressure and heart rate normalized, and their parasympathetic system helped the body recover.[6]

Li also found that when people spend about two hours a day walking in the forest, their sleep patterns improve dramatically. We know cortisol and other stress hormones can cause sleep deficiencies and lead to many severe health problems. Li measured sleep activity of patients using a sleep polygraph and accelerator, testing patients before and after forest bathing. Their sleep time increased an average of about one and a half hours. And in other sleep studies he showed patients' sleep average increased by forty-five minutes, and patients reported that they had less anxiety.[7]

Li's research determined that forest bathing enhanced the mood of participants. He had them take the Profile of Mood States test before and after forest bathing. This test measures emotions like sadness, terror, guilt, and confusion, as well as states like vigor and exhaustion. All walking helped, but forest walking in particular showed increased vigor and decreased exhaustion, particularly in women.[8]

These findings are not entirely surprising. Most people live in urban environments. Their workdays are long and chaotic. Their commutes are difficult and tiring. They build up a lot of residual stress in their systems. It is therefore logical that spending hours in a beautiful natural setting, whether it is a forest, a city park, or a garden, is beneficial. But the most interesting part of this research is the finding that spending time in nature dramatically strengthens your immune system. Stress does inhibit the immune system, but forests do more than help us reduce stress.

Natural killer (NK) cells are a type of white blood cell that attacks and kills viruses, tumor cells, and other unwanted cells. NK cells contain anti-cancer proteins such as perforin, granulysin, and granzymes.

These proteins perforate the cell membranes of unwanted cells and deliberately kill the cells they target. Higher NK activity lowers the risk of cancer and other diseases. Studies by Li found that forest bathing increased NK activity by fifty-three percent and NK cell numbers by fifty percent. The anti-cancer protein granulysin increased by forty-eight percent, granzyme by thirty-nine percent, and perforin by twenty-eight percent.[9]

When he later tested participants, he found that the results lasted for as long as thirty days. He concluded that one forest-bathing trip could maintain a high level of NK cell activity for a month. In later studies, he also determined that people who live in areas with fewer trees have higher stress levels and higher mortality rates.[10]

These research findings validate what many of us already believe. And there is extensive anecdotal evidence behind these scientific conclusions. If you were to ask Li's subjects about their experiences, they would say things like, "I didn't really believe forest bathing could make me feel better. I was surprised at how effective it is. The feeling lasts all night," and, "it is similar to the way doing sport can relieve stress . . . now that I have experienced it myself I know how powerful it can be."[11] And this story from a writer who suffers from writer's block: "I have a place I always go. It's full of wild thyme and rosemary and often I just stand and breathe. As my eyes travel the landscape I can almost feel my brain untangling. Sometimes it is as though the answer I'm looking for is right there in the trees and all I had to do was get there."[12] We love being in forests, at the ocean, and in the mountains. We can't get enough. We keep going back for more. We do it again and again because it makes us feel better. Now we have scientific findings that back up what we already knew.

Just how do forests enhance our immune systems so dramatically? We know these results are not solely due to stress reduction or an increase in available oxygen. Something else is going on, and what it is may surprise you. It has to do with one of the many ways that trees communicate. Remember that trees communicate in part through smell—that is, by emitting scents. These scents are emitted as defenses and warnings. They are known generally as phytoncides.

South Korean scientists found that elderly women who regularly walk in forests show improved blood pressure rates, better lung capacity, and improvement in the elasticity of their arteries. These findings did not occur among women who walked only in urban areas. These scientists believe that the phytoncides emitted by trees kill germs that would otherwise affect our immune systems.[13]

The Korean Forest Agency now offers degrees in forest healing. There are three official healing forests in South Korea and dozens more are opening soon. These healing forests are frequented by cancer patients and allergy sufferers and by all those who simply love the forest. South Korea is taking the lead in promoting its amazing forests as a source of healing. Their Forest Agency is building a $100 million forest healing complex, which will even have an addiction center. They are also distilling oils from forest trees to determine their effects on allergies and on killing bacteria.[14]

The forest air is full of phytoncides. They are the natural oils or scents that trees release to defend themselves from bacteria, insects, and predators. Evergreens produce the most phytoncides. Their main components are organic compounds called terpenes. Terpenes are what you smell when you walk in a conifer forest. The main ones are D-limonene, which smells like lemon; alpha-pinene, which smells like pine; beta-pinene, which smells like dill or basil; and camphene, which smells like turpentine.[15]

When Li tested human NK cells with phytoncides in lab studies, he found that NK cell activity and the presence of anti-cancer proteins perforin, granzyme A, and granulysin all increased. In human subjects who were exposed to phytoncides, he found that all had significantly lowered stress hormones, slept better, and had decreased tension, anxiety, anger, and fatigue. Other researchers showed that phytoncides stimulated pleasant moods, lowered blood pressure and heart rate, and suppressed the sympathetic nervous system.[16] A study at the Mie University in Japan found that the phytoncide D-limonene was more effective than antidepressants.[17]

In case you are concluding that most of the research that equates walking in the forest with health benefits was conducted years ago by a few Asian tree huggers, I would like to acquaint you with a couple of comprehensive research summaries. A major review of field experiments on forest breathing was published on May 16, 2019 by *Global Advances in Integrative Medicine and Health*. This review article focuses on recent research citing the beneficial effects of forest bathing on heart rate variability, expressed as an increase in a factor called InNF, indicating activation of the parasympathetic system and its effect on reducing anxiety. It reviewed numerous research studies with hundreds of participants and found "strong and consistent evidence that exposure to forest bathing results in an increase in factors associated with activation of the parasympathetic nervous system and also reduced anxiety. Additional therapeutic benefits include positive mood states, improved mental coordination, with reductions in stress levels and lower blood pressure."[18]

A meta-analysis was published in *Environmental Research* in October of 2018 by researchers from the University of Anglia in the UK. It linked time spent in nature with health benefits. They reviewed 103 observational studies and forty interventional studies. They tracked 290 million participants from twenty different countries. Alice G. Walton summarized the result in a Forbes.com article in July of 2018. Walton reported that these studies showed that spending time in green spaces was linked to reduced cortisol levels, lower heart rates, less risk of heart disease, lower blood pressure and lower cholesterol, decreased risk of developing type II diabetes, reduced mortality, and reduced death from heart disease. For women who were pregnant, spending time in green spaces also reduced the risk of preterm birth and smaller babies. Studies also showed links between green spaces and cancer outcomes, sleep issues, and neurological conditions.[19]

This meta-analysis, entitled "The health benefits of the great outdoors: a systematic review and meta-analysis of green space exposure and health outcome," only considered articles published in English. It also showed increased incidence of good self–reported health and a reduction in strokes, hypertension, dyslipidemia, and asthma. Green space was defined as open, undeveloped land with natural vegetation, including urban parks and open spaces. It was not limited to forests.[20]

There is a common species of bacteria that is found in soil called *Mycobacterium vaccae.* It has been found to result in positive feelings, higher energy levels, and better cognitive functioning. Scientists have found that the neurons *M. vaccae* activates are associated with the immune system. Exposure to soil stimulates the immune system, making us feel happier.[21] I think most gardeners would agree. In fact, I have long wondered why I feel so good when I am gardening and digging in the soil. Most of us can't wait until spring. We have to get our hands in the soil. I thought it was just because I grew up on a farm and we grew much of our own food. I believed it was just a habit. It turns out it may be a type of addiction.

A recent University of Colorado Boulder study showed that injecting this soil bacteria in mice made them more resistant to stress. Other studies have shown that *M. vaccae* increases serotonin in the prefrontal cortex and can help prevent PTSD-like symptoms in mice.[22] When we are actively gardening, we may be increasing our serotonin levels in a most natural way. Long, cold winter days may have repressed those levels. Perhaps this is a natural seasonal healing process. If so, nature's fix is far better than any antidepressant. Twelve percent of our population takes antidepressants. It's time we all consider getting down on our knees as gardeners do and dig around a bit. Let's get dirtier in the out-of-doors. It may really be good for us.

Many countries have cultural traditions that involve walking in forests. On my first visit to Europe, I visited Bavaria and hiked the beautiful trails of the Bavarian Alps. I was not alone. Bavarians love their mountains. I hiked with families and with solitary hikers, many dressed in their beautiful wool felt trachten jackets with deer horn buttons, Tyrolean hats, and ornate walking sticks. They made quite a fashion statement, but their style of dress, though quaint to me, I knew was a source of pride to them, as were their well-groomed and often paved forest trails. Forest walking was just what everyone did.

Finland also has a tradition of spending time in forests. Finland is densely covered with trees. Writer Florence Williams points out in her book *The Nature Fix* that until recently, much of Finland was rural and even its urban areas are still close to large tracts of forest. So, most urban Finns still have access to farms and woodlots. Williams cites Finnish studies that show health care costs can be reduced when people spend time in nature. These studies also look at work benefit measures like job happiness, productivity, and creativity. Researchers found that the biggest boosts occurred when a participant spent at least five hours a month in nature. The average Finn spends time in nature several times a week. This includes berry picking as well as skiing and skating. Outdoor recreation is a year-round passion. As Williams suggests, this may be why Finns score the highest on the global scale of happiness.[23]

In the United States, we may not have these cultural traditions, but we love our parks. We have sixty-three national parks and they are visited by over 330 million people a year. They may be loved to death, in fact. And outdoor recreation is big business in the US. According to the Department of Commerce, our love of nature translates into about two percent of the economy, some 422 billion dollars and some 4.5 million jobs.

In my early years as an environmental educator, outdoor education was just catching on. An environmental college I was affiliated with developed one of the first outdoor education programs in the country. Of course, students flocked to it. You could actually earn college credits for spending time outdoors. No one quantified the effects of spending time in the wild, but we knew it made us feel good and, for some of us, it was addictive. Much later, programs were developed that demonstrated how effective time in nature was for conditions like PSTD, juvenile delinquency, and attention deficit disorder.

According to the materials provided by the US Forest Service Pacific Northwest Research Station, published in 2010, studies show that after stressful situations or ones that demand concentration, people recover better in natural environments. We now know why.

Florence Williams, author of *The Nature Fix,* explains what neurologists are now finding about the effect of nature on cognitive functioning. It turns out that time in nature can make us more productive. We know regular exercise can help with cognitive decline. That has been well documented. But why does immersion in nature also help our productivity? It has to do with attention and how our brain works. We are very easily distracted. Staying focused is hard. And now we are living in a digital world which has increased our distraction level exponentially. We are guinea pigs in an uncharted experiment. We don't really know if what our brains are gaining in terms of more information and memory storage will be worth it. There is a tremendous cost involved. We are exhausting our brains. Our brains filter a lot as it is. We have always had more information than we can deal with. So, we developed selective attention. Our brains turn a lot off so we can pay attention to what is the most important. It's a survival skill. Sadly, our brains are actually extremely slow at processing information. Stanford neurologist David Levitin tells us our processing speed is only about 120 bits per second, and it takes 60 bits a second for us to understand one person speaking to us.[24] Our attention is a limited resource. We are easily overloaded. We make mistakes, get irritated, and waste the energy (glucose) we need for clear cognitive functioning. Time in nature helps us recover. Our brains can rest in the beauty of a sunset, the sight of a blue heron rising from a marsh, the calm repetition of the ocean surf.

There are far more basic ways that nature heals us. Some 60,000 years ago, early humans and our Neanderthal cousins recognized that certain plants had healing properties. Large amounts of pollen from numerous plant species that are still used as herbal remedies have been found near their ancient dwelling sites in northern Iraq. Early human sites have vestiges of herbal plants like nettle and chickweed.[25] This information about healing plants passed from healer to healer for thousands of years.

Even animals know about healing plants. Yes, animals self-medicate too. We know dogs eat grass to induce vomiting. Baboons eat certain plant leaves to combat flatworms. Lizards eat a certain root to counter venom from snake bites. Macaws eat clay to help with digestion and kill bacteria. Pregnant elephants eat certain tree leaves to induce delivery.[26]

We know that Sumerians used hundreds of plant medicines. References to these early medicines, including opium, have been found in clay tablets some 3,000 years old. They used a form of writing called cuneiform, which is the oldest form of writing that we know of. Scribes used a reed stylus to make indentations on wet clay slabs that were then baked. It only made sense that they would record information on lifesaving medical plants. Egyptians in 1550 BCE used papyrus to document important plant remedies. In fact, most early civilizations documented the therapeutic use of plants in one way or another. These traditions evolved into homeopathic, ayurvedic, Chinese, and other plant-based systems of medicinal treatment.[27]

Many commonly prescribed drugs are derived from nature. The anticancer drug Taxol comes from the bark of yew trees. Common plant-derived drugs include aspirin, digoxin, quinine, and opium. Many are from angiosperms—flowering plants. There are four main categories of plants commonly used pharmacologically: alkaloids, glycosides, polyphenols, and terpenes. Alkaloids are bitter, occasionally toxic, and include atropine, a nightshade, which is used to treat heart rhythm problems and stomach and bowel issues. Berberine, found in goldenseal and other plants, is used for diabetes, high cholesterol, and high blood pressure.

Other alkaloids give us coffee, cocaine, ephedrine, morphine, nicotine, and quinine. Glycosides are found in plants like rhubarb, senna, and aloe, and can be used as laxatives. Some glycosides like digoxin and digitoxin come from foxglove and lily of the valley—they are heart medicines and act as diuretics. Digoxin is used for atrial fibrillation and even heart failure. Polyphenols include tannins and the phytoestrogens used for gynecological disorders. This category includes plants like kudzu, angelica, fennel, and anise. Then there are terpenes, which, as we have noted, are often found in coniferous plants. These resinous compounds have many medical uses. For example, thymol, which is found in the common herb thyme, is used as an antiseptic. Terpenes are often used in aromatherapy.[28, 29]

Medicinal plants are still widely used across the world. The Royal Botanical Gardens, Kew, estimates that over 17,810 plant species are regularly used medicinally. Many of these critical plants are now threatened by climate change, habitat loss due to agriculture use, and overharvesting.

Merely looking at plants through a window can lead to healing. In the 1980s, an American professor named Roger Ulrich collected information on patients who had just had abdominal surgery and found that those who recovered the fastest were in hospital rooms that looked out at trees. Later, in a Swedish hospital, he found that patients who were shown pictures of trees after heart surgery recovered faster.[30]

The science is conclusive. Forests and all of nature are healing to humans. All we have to do is show up. We can sit, we can walk, or we can lie down. It all works. Clemens G. Arvay has written a book called *The Biophilia Effect,* which is full of exercises one can do in the spirit of forest bathing. Arvay points out that the concentration of anti-cancer terpenes in the forest air varies seasonally. They are highest in the summer and lowest in the winter, with a peak in July and August. Most are found in the middle of the forest rather than at the edges. They are most abundant after a rain when the air is moist. They tend to be the densest near the ground, at our level.[31]

And if you don't have a forest nearby, there is plenty of research indicating that any time in nature, whether it is a city park, a stream, a walk on the beach, or just time in your garden, will do. And remember to get your hands dirty so you up those serotonin levels.

The term biophilia was first used by Erich Fromm, a German-born American psychoanalyst. Biophilia means "love of life or living systems." Fromm used the term to describe a psychological orientation of being attracted to all that is alive and vital.[32] Later, in 1984, Edward O. Wilson popularized the term in his book *Biophilia*. He defined it as "the urge to affiliate with other forms of life," suggesting that it describes the "connections that human beings subconsciously seek with the rest of life." He suggested that humans have a deep positive connection with all life forms and nature.[33]

The concept of biophilia implies that we are not separate from the rest of nature, that we are a part of all life. We have always been a part of nature and are implicitly a part of the great web of life. We did not evolve in petri dishes or concrete boxes. We walked forest and savannah paths filled with *M. vaccae,* not asphalt. The air we breathed was full of terpenes, not hydrocarbons. The ground we slept on, in all likelihood, had naturally occurring serotonin.

Presently, about fifty-five percent of us live in urban areas. This will increase to two-thirds by 2050. The average American spends seven percent of their time outdoors. Worldwide, the average person spends eighty-seven percent of their time inside. In 2019 the Velex Group surveyed some 17,000 adults across fifteen countries. According to their Indoor Climate Expert, Peter Foldbjerg, the average person spends ninety percent of their time indoors. He laments that "We are becoming an indoor generation." Sadly, most children spend even less time outdoors than adults do.[34]

Science tells us that nature can help us de-stress, strengthen our immune systems, and live happier lives. We don't need to rely on an insurance card, a prescription, or our doctor's advice. If we are interested in using nature's readily available and free biomedical assistance, all we have to do is show up. But that seems to be the hard part.

The effects of trees on our physical health are now understood and accepted. Nature's effects on our psychological health are also well documented. There is no question that time spent in nature helps reduce the impact of our harried and stress-filled lives.

But there is more to how nature heals us than terpenes, immune effects, brain relaxation, and herbal remedies. Sapiens have been around for some 177,000 years. Up until about 10,000 years ago, we lived in small groups. We had well-honed hunter-gatherer skills and were animistic. We believed we were part of nature and that the natural world around us was vibrant and imbued with consciousness. We were both predator and prey, but according to Yuval Noah Harari, the author of *Sapiens,* life was not that bad.

As animistic creatures of the forest and savannah, we understood ourselves to be interconnected with all other species. We were part of a unified whole and celebrated our connection to all of life through rituals, songs, stories, and dance. We do not have the details of their lives, but from the limited artifacts and cave paintings available, we can conclude that our ancestors celebrated nature and knew it as source of awe and wonder. Of course, many aspects of their lives were challenging. There were dangers everywhere. Infant mortality was extremely high. But their world was still enchanted.

Harari calls the agricultural revolution, which occurred about 10,000 years ago, "history's biggest fraud." Humans began to devote all their time to domesticating a few plants and animals. It began in southwestern Turkey and western Iran. Wheat and goats were domesticated 11,000 years ago. This was followed by peas and lentils, then olives, then horses, and then grapes some 5,500 years ago.[35] Harari insists this did not make us smarter or result in greater well-being. It led to a worse diet and much less leisure time. Early agriculture required grueling dusk-to-dawn labor. Early farmers suffered from malnutrition and disease. Large groups often lived in crowded, unhealthy conditions.[36]

This innovation from hunting and gathering to agriculture led to larger, cramped villages. By 2900 BCE, the Mesopotamian city of Uruk had over 80,000 people. Humans no longer saw themselves as part of nature. Leisure only existed among the ruling and priestly hierarchy.

We estimate the total global human population in 10,000 BCE was about a million. By 2800 BCE, it was fourteen million. And fast forward 5,000 years, it is 7.7 billion. In our era, most humans live in megacities that each have populations of more than ten million people. They spend little or no time in nature. Their lives are far removed from their animistic ancestors.

We now get most of our food from a half dozen or so angiosperms. All cereal grains, including rice, barley, wheat, millet, and corn, come from flowering grasses. They evolved in the Cretaceous period, after the dinosaurs. We can't digest most grasses, but we can eat their fruits, especially if they are cooked. The grains we eat are the fruit of these grasses. These were the crops that allowed human population to expand so rapidly and take over so much of the Earth.

In addition to grasses, we eat a few other plants, but not that many. In all there are some 437 plant families in the world. We dine on legumes, brassicas like cabbage, nightshades such as tomatoes and potatoes, members of the parsley family like carrots and celery, and family group members like pumpkin and melon. Most the fruit we eat comes from the rose and the citrus families. The rose family includes apples. The palm family gives us palm oil and coconut.[37] Most of the plants that give us sources of food are grown in monocultures.

Overall, plants make up eighty-two percent of the Earth's biomass, bacteria make up thirteen percent, and of the remaining five percent, humans make up a mere one percent. But what a mess that one percent has made. The domestication of animals by humans is a big part of the problem. Consider that seventy percent of all birds are domesticated. That includes some twenty-two billion chickens. Only four percent of mammals are now wild.[38] Presently, we use about thirty-eight percent of the Earth for food production. The remainder is not suitable, is forest land (much of which is planted in monocultures), or is used for urban development. Seventy-seven percent of all agricultural land is used for livestock. There is not much land left for us to exploit or to save.

This is not sustainable. We seem to understand there is a problem, but we have been thus far unable or unwilling to change. As we have discussed, modern life and the modern myths we have created about ourselves have left us confused and often ignorant of reality. Many suffer from the three Ds: dis-eased, disillusioned, and disassociated. These conditions and their resultant cognitive biases, if not balanced by our inherent values (including valuing the survival of our fellow species), act like parasitic viruses that sap our goodness and creativity.

No one is suggesting we entirely eliminate our separate sense of self, our subject-object perception, or dualistic reasoning. We know these human attributes reinforce our sense of separation from all other species and can make us crazy. But they also define us and, in many ways, we need them to survive. It is useful to think of these perceptual attributes as tools that require skillful use. Like any tools, they need to be used thoughtfully and in a balanced manner.

What is required is that we reaffirm our sense of who we are. That is, reaffirm our understanding of ourselves as a unique, multitalented species within a larger whole. We need to re-envision our world as holistic and reunify our consciousness with the larger whole we are an implicit part of. That is not nearly as difficult as it sounds. It is merely a perceptual shift. All of nature is available to help us make this transition to health and wholeness.

De-stressing by spending time in nature is a critical first step, but there are many levels of nature's healing repertoire that help balance our bodies, minds, and souls. Biophilia, love of nature and all life, is wired into our DNA. It is a natural process. We don't have to resume archaic animistic rituals. We do not have to do anything. All we have to do is show up. When we place ourselves in a beautiful natural setting, we become ourselves. We automatically let go of our daily stresses. When we spend enough time in nature, we slowly quiet our minds; we become more inwardly still and observant. Then the magic begins. We begin to perceive differently. We become more aware of all of the life around us. We become aware we are immersed in something larger and far grander than ourselves—something we are an intrinsic part of.

There are three main networks in our brains. Most of the time the executive network is the one that is engaged. It is the task-oriented prefrontal cortex. This may be what many of us think of as our minds. It supports cognitive behaviors, emotions, and thinking. There is also the spatial network, which orients us in space. The third network is the default-mode network. This is where daydreaming occurs. It may also have to do with insight, creativity, and empathy. When it takes over, the executive network slows down—it can rest. It may take several days in nature to truly rest the executive network. But then, researchers say, something else happens. You get into the flow, like athletes and artists do. Everything is more beautiful and in balance. You are relaxed and you notice new details. The default network is triggered by nature.[39] This is something we have all experienced, whether we are aware of it or not. It can happen just by walking in a park, along a steam, or gazing at the clouds floating in the sky. It is effortless. Suddenly, your perception changes.

Scientists believe the executive network contributes to our sense of self. They have found that it is naturally suppressed when a subject is meditating. It is suppressed because it receives less blood flow and oxygen when one meditates. A meditator can have normal awareness of sights, sounds, and thoughts but no sense of self as the thinker of thought and perceiver of sensory input. This implies the self is merely a construction. It can be minimized when it is not really needed. Researchers have also studied individuals under the influence of psychedelic drugs and monitored their related brain activity through functional MRI scans. They found drugs like LSD, ketamine, and psilocybin also quiet the circuit in the brain that connects the parahippocampus and the retrosplenial cortex in the default-mode network. Just as in meditation, at such times, consciousness is present, but there is an experience of the loss of the sense of self. People report experiences of wholeness, nonduality, and deep inner peace.[40]

Psychedelics also lessen the communication among neurons in other areas than the default-mode network. This makes brain activity less segregated. Robin Carhart-Harris, one of the top researchers in this field, states: "the separateness of these networks breaks down and instead you see a more integrated or unified brain. . . . The barriers between the sense of self and the feeling of interconnection with one's environment appear to dissolve."[41]

These neurological findings explain what can happen in deep meditation and mystical states. They also explain what happens when nature helps us drop our sense of self. We begin to perceive in a more holistic, animistic manner. This is what happened to me when I was a child playing in the woods. Everything was alive and I was a part of it. As a young environmentalist immersed in the forests, streams, and lakes of the upper Midwest, I implicitly understood the holistic nature of the world I was a part of. Now, as an elder lover of all alpine forests and valleys, I can easily access the deeper states of connection that the natural world is a gateway to. It happens in an instant. The provocateur may be the sound of a red-winged blackbird, a croaking leopard frog, a shift in the breeze, and suddenly my heart explodes. In that instant, a sense of becoming one with everything overwhelms me. There is no separate self in these moments. The experience is as true and direct as an arrow shot from a bow. I am suddenly and simply one with all of life.

When nature's gateway opens us to the wonder of all existence, the mind as we know it stops; we drop all sense of self and are simply in the moment—a moment filled with exquisite beauty and deep connection. Whatever remains of what I think of as "me" is totally enraptured in awe and love.

Some environments are more evocative than others. Breathtaking beauty and other deeply aesthetic experiences are provocateurs that engage the default network. Turn off your cell phone and walk in the redwoods. You are almost guaranteed a deep encounter with the luminous. The key here is turning off the distractions and just being present.

If you do not have a redwood forest to lose yourself in, find a tree to hug. Yes, I am a tree hugger. Sometimes while in nature, I am filled with so much joy I just have to hug a tree. I can't help myself. I especially like hugging large ponderosa pines. They are so fragrant, and smell like vanilla and butterscotch.

There is a retreat center in northern New Mexico called Vallecitos. It is a breathtaking place—a bit like a small Yosemite. It has several old-growth ponderosa pines. There is one giant specimen near the trail that I am sure gets dozens of hugs a day during any given retreat. If you sit and watch people walk by, you will see that they can't help themselves. They come up to the tree; they circle around it, check to see if anyone is looking, and then before you know it they are in a deep and sometimes tearful embrace. There is something profound that happens. There is an exchange of loving feeling—of connection. I don't know why. Maybe it is like petting your dog. We know that petting your dog is therapeutic to both you and your dog. Part of it is just the touching and bonding experience, but it is also healing because the touching releases oxytocin, the neurohormone that makes us feel good, loving, and cuddly. It's okay. It is good for you. You can be a silent tree hugger. No one has to see you do it.

This loving, empathic feeling may be connected to the vagus nerve, which some researchers say is the seat of empathy in humans. We know it affects the parasympathetic system, which slows our heart beat and helps us come back into balance after fear-based responses or when we have anxiety. In addition to having a role in generating loving and empathic responses, the vagus nerve is connected to our oxytocin receptors. This nerve also seems to respond to awe. Nature fills us with awe.[42] And you can engage the vagus nerve through simple things like slow belly breathing, singing, or chanting. I would add tree hugging to that list.

The natural world offers an extraordinary opportunity for us to connect with our true, whole selves and to grasp, even if momentarily, the unity of all existence. For many of us, this is the easiest and most accessible path to our innate consciousness. This is the path to our true self, our natural self, the same self that all spiritual paths lead to. We learn that this is what it feels like to be truly human.

Nature has the capacity to ground us in the ground of all being. It is a value of such extraordinary importance that it is beyond any human measure. Without it, we are shadows of ourselves and cannot grasp our true humanness. You don't have to sign up, take vows to any religion, or bow to other humans. This blessing requires a different form of supplication. You don't have to let go of anything but your small, illusionary, separate self and you merge with the larger whole. It is entirely joyful and ecstatic.

The great nature writers John Muir, Aldo Leopold, Wendell Berry, and countless others eloquently articulated this extraordinary power inherent in nature. Saint Francis accessed this ecstasy in his garden. Hildegard von Bingen called it *viriditas,* the greening power of the Divine.

It now has another name: *eutierria*. *Eu* is the Greek word for good. *Tierra* means earth. Eutierria is defined by Australian sustainability professor Glenn Albrecht as "a good and positive feeling of oneness with the earth and its life forces." He says it arises when "the human-nature relationship is spontaneous and mutually enriching." Environmental writer Kenneth Worthy equates it with the "oceanic" feeling mystics of many religious backgrounds describe. Worthy says "when it occurs your perception of the boundaries between yourself and all else—the thoughts and feelings setting you off from the rest of the cosmos—seem to evaporate." When this occurs, he explains, you become one with the universe and a sense of harmony and connection infuses your consciousness.[43]

A walk in a forest also fills us with a sense of gratitude. We are given so much by nature—the oxygen we breathe and the food we eat—and through its generosity we are healed in body, spirit, and soul. This awareness alone opens our hearts, resulting in even deeper healing. As we shed layers of culturally accumulated misunderstanding and confusion, our true self emerges. That is nature's special gift to humanity. It helps us reaffirm our identity. With blinding light, it shows us that we, in our singular brilliance, are an implicit part of the whole. It whispers that we have never left its embrace.

4
OUR TREE CONNECTIONS

Linden (Tilia)

In Germany . . . it was the traditional hub of village life where people would meet or sit on benches in its shade. . . . It also constituted the central point in village feasts. . . . However, a more serious task for the linden was to be the location for the local court of law. This custom dates back to pre-Christian times, when tribal gatherings were held underneath sacred trees. It is revealing that the ancients gathered, discussed and judged underneath, the 'female' linden which represents mercy.

The Meaning of Trees

It may come as a surprise, but we actually are like trees in some key respects. Of course, the inability to photosynthesize remains, even though humans have tried to emulate the ability that our green friends have mastered. Some have even insisted they have attained this special skill. There are accounts from various religious traditions of individuals who seemed to photosynthesize. That is, they believed that they simply absorbed energy from sunlight and did not take in nutrients. One such individual was a German Catholic mystic named Therese Neumann.

Neumann said she had consumed no food other than the Holy Eucharist or drunk any water from 1926 until her death in 1962. Her claims were found to be suspicious by local medical authorities. But Paramahansa Yogananda, the highly respected author of *The Autobiography of a Yogi,* devoted a whole chapter in his book to her.[1]

The term "breatharianism" describes the belief that it is possible for a person to live without consuming food and, in some cases, water. In Hindu traditions, some believe that humans can be sustained by sunlight alone, which is one of the main sources of prana or vital energy. Prahlad Jani is an Indian sadhu who says he lived without food and water for more than seventy years.[2] Unfortunately, some practitioners of breatharianism, also known as Inedia, have died of starvation or dehydration. Obviously, it is not an easy or risk-free lifestyle.

While breatharianism is not anything most of us will try soon, scientists are studying animal organisms that have developed the ability to work symbiotically with photosynthetic chloroplasts. Certain sea slugs appear to be powered by the sun. One mollusk slug species has been found to have a symbiotic relationship with chloroplasts. They actually steal the chloroplasts from algae. As a result of their enterprising behavior, they have been found to survive up to nine months on just light and carbon dioxide.[3] That does not mean that humans are likely to develop the ability to photosynthesize at some point; it would require considerable genetic modification, so it is highly unlikely we will have green skin anytime soon.

There are many characteristics that we do share with trees, however. As we have seen, the plant and animal kingdoms have evolved very differently. But we share many attributes. Species in both kingdoms appear to demonstrate social behaviors, like cooperation, and are guided by hormonal drivers. They have both similar and very different ways of sensing their environment and gathering information that enhances their likelihood of survival. They can learn, and they demonstrate a type of intelligence or smartness as they navigate their diverse environments.

Cooperation is found throughout all three kingdoms of life. The cells of animals, plants, and fungi all contain mitochondria. The mitochondria in all three kingdoms descended from early eukaryotic cells—the first evolutionary example of cooperation.

As we have seen, trees seem to be cooperative. They live in expressive interactive networks where they share resources, information, and defenses in a manner that may make human cooperation seem impoverished. A forest seems more of a collective entity than a system made of individuals. I am no longer sure how to think of trees. They seemingly share scarce resources with other species in situations where they have no direct benefit. By our terms, that would be kindness or altruism. We don't really have a way to understand these behaviors and what they might actually mean in the plant world. It has been difficult enough for us to understand the origin of human forms of cooperation and altruism. For too long, we mistakenly assumed that it was primarily competition that ruled human behavior.

Darwin's theory of natural selection, from which the idea of the "survival of the fittest" and the competitive nature of humans is derived, has held sway in our thinking for far too long. This theory has overshadowed the fact that humans are among the most cooperative mammals, not the most competitive.

Scientists have long struggled to understand cooperation and altruism and how it originated. They have struggled to reconcile it with Darwinism. How did humans become so cooperative and why? After all, natural selection implies that self-interest leads to success. But Darwin himself acknowledged that humans were also cooperative and was fascinated by the cooperation of social insects. Gradually, with some delicate stitching by evolutionary theorists, cooperative behaviors were recognized as being genetically wired into our species. But it was not until 1964 that William Hamilton popularized "kin selection," the idea that we sacrifice for others when we are related to them. A kin selection gene could spread cooperation if the cooperative individual survives and reproduces.[4]

In 1975, Harvard biologist E. O. Wilson, without contradicting Darwin, and relying on kin selection, published a widely accepted theory that stated that social behaviors, including altruism, are genetically programmed to help a species survive. In 2010, Wilson changed his mind. He said altruism emerges to protect social groups whether they are kin or not. He said this is what drives evolution, not just self-sacrifice to protect a relative's genes.[5] In other words, it is the community or group that is important and what ultimately shaped our behavior and evolution, not the individual. That was a big leap.

Human cooperation is, of course, different than cooperation among animals, much less plants. Social insect societies, like bees and ants, have high levels of cooperation and are genetic siblings to their hive mates. They are all kin. Human societies also have extensive cooperation with unrelated individuals and complete strangers, not just with kin. We may receive direct benefits, indirect benefits, or no benefits at all for cooperation. Human cooperation may be reciprocal and can be maintained by reward, punishment, ostracism, and reputation-building, as well as by far more complex motivitations.[6]

Researchers Alicia P. Melis and Dirk Semmann point out that humans have unique cognitive abilities that allow them to remember past interactions, track collaborative behaviors of others, and transfer this information to others. This creates a complex social network. Language may make a huge difference in explaining our greater ability to cooperate and enforce cooperation. It is through language that human societies and groups create complex laws, morals, and traditions, and maintain social rules.[7] As Harari, the author of *Sapiens* emphasized, enhanced language skills were a part of the cognitive revolution and allowed for cooperation between larger and larger groups based on the myths and stories they shared.

We know that humans can be greedy and competitive or kind and generous. We like to believe that we are moving the fulcrum toward the kind and generous part of the spectrum. We don't yet fully know to what degree these behaviors are shaped by our genes. Ariel Knafa, a professor of psychology at Hebrew University, suggests cooperation may be innate, as infants show empathy for others in distress. She says that in studying twins, they found overwhelming evidence of a genetic component to sharing and empathy.[8] Empathy appears to have a neurological base and is associated with the vagus nerve and oxytocin receptors.

Knafa also mentions the theory that evolutionary processes take place at the group level, not the individual level. Highly cooperative individuals have a higher chance of survival, as they can work together to reach common goals. Again, it may be that the social group is the actual driver of evolution, and therefore behaviors that lead to group cohesion are the ones that are passed on. Maybe, from an evolutionary perspective, the group is more important than the individual. Could we be evolving a collective form of intelligence that shapes our cooperative behavior? Certainly, the digital age is opening our understanding to new ways to look at group behavior and decision-making.

Of course, there is more to the story than genetic predispositions. Unlike reptiles, mammals are immature at birth. Mother sea turtles lay their eggs in the sand and return to the sea, never to see their offspring again, but mother mammals have the brain and hormonal circuitry that compels them to provide for their helpless young. Both they and their newborns feel pleasure when they are near each other. This feeling of pleasure extends to children, mates, kin, friends, and pets. This good feeling is related to the release of the neurohormone oxytocin.[9] This is only one of several hormones that influence caring behaviors. And we are not the only creatures who like the cuddly feeling that oxytocin gives. Even in fish and reptiles, a type of oxytocin influences reproduction behaviors like egg laying and spawning.[10] And given that we now know that even dinosaurs were social animals and took care of their young, it is possible oxytocin directed their behavior to some degree. Who would have guessed they could also be cuddly?

The release of the neurohormone oxytocin results in mother–baby attachment, which gives rise to love and caring. We know that oxytocin stimulates complex social behaviors, including maternal behavior, like lactation and licking of offspring. It apparently also helps mothers recognize the smell of their children. Researchers have found that there are genetic differences in people's responses to oxytocin. These differences are linked to an individual's ability to read faces, infer another's emotions, and feel distress at another's hardship.[11] We also now know that in humans, oxytocin's role is spread far beyond mothering behaviors. A study in Zurich by Ernst Fehr showed that people exposed to oxytocin were more trusting of strangers with their money than those who were not.[12] We may not be happy to hear that the world of economics and finance has caught on to this trust-inducing hormone, but it has.

Oxytocin travels through the bloodstream and can affect organs far from the brain. It is like a neurotransmitter in that it allows brain cells to communicate, but unlike dopamine and serotonin, which have numerous receptors, it delivers its signal through one receptor. This receptor is a protein designed to recognize the shape of the hormone, and the receptor's ability differs in individuals, implying some are naturally more responsive to the hormone.[13]

We also get positive feelings from other neurochemicals like cannabinoids, a type of opioid. They are produced in the brain and reward social interactions. They can be released when social animals like us are with friends or family. They are the source of the pleasure we take in socializing.[14]

The pandemic has demonstrated how strong our needs are for social contact with others. It comes as no surprise that the use of cannabis products has also increased during a time of required social distancing and repeated lockdowns.

In short, in highly social mammals, oxytocin is released in the brain in social situations like food-sharing, cuddling, sex, and grooming. Cannabinoids also intensify social interactions, increasing cooperation and trust. This is how social animals become sharing, caring, cooperative animals. There may be a genetic propensity that makes some humans more trusting and more kind than others. Some humans may have more functional receptors to hormones like oxytocin, or they may have more receptivity because they are innately trusting and kind.

As social animals, we take pleasure in being with others and in cooperating and caring and sharing with others. Neurochemicals like oxytocin and cannabinoids encourage group interaction and network behaviors. They encourage kindness and altruism. Although some modern theories on human cooperation suggest that acting selflessly provides a selective advantage, it is pretty clear that early humans also cooperated merely to survive. They needed to work together to obtain food and to defend themselves. Mutual cooperation was based on interdependence. Hunter-gatherers had to forage together. Because they were by nature and necessity interdependent, special cooperative abilities evolved. They learned to divide their food equitably, to communicate goals, and to understand that as a community all their roles were important.[15] They learned to share resources and to trust in the community as a whole. That tells us they learned to function as an interactive network. Their survival was dependent on it. The key takeaway is that early humans were an interdependent community.

There has been a lot written on kindness, altruism, and empathy in recent years. A wealth of research studies imply that these characteristics may be innate in humans. Several of these studies tell us that we can increase our happiness through practicing kindness. Lee Rowland is an experimental psychologist and researcher who studies kindness, cooperation, and behavioral change. His research summaries tell us that kindness reduces anxiety, that altruistic individuals are held in higher esteem, that empathy can even reduce the severity and duration of the common cold, and that spending money on others is good for your heart, by reducing high blood pressure more effectively than anti-hypertension medications.[16]

Rowland cautions that some of these are single studies and not conclusive, and they vary in the way kindness is treated.[17] So, along with his colleagues, he published a meta-analysis on the research on kindness and found that practicing kindness did indeed significantly increase well-being.[18]

Many people, particularly Buddhists, practice loving-kindness meditation and other compassion-based practices. Some studies have shown that these actions increase the well-being of the practitioners.[19] Most of us know from our own experience that helping others simply makes us feel better. But that does not imply that we help others out of self-interest.

Helping other humans and other species improves our well-being. These behaviors reduce anxiety, bolster our immune system, and decrease blood pressure. And being loving and kind to others just makes us feel good. We like to help others. We like to make things better for everyone. That is just our nature, which means we have natural empathy for all of life. If we want to have a scientific reason for this, I think we need to look no further than our own oxytocin and cannabinoid receptors. Generosity, sharing, and connecting are, in many ways, what we are about. Of course, we feel good when others feel good. At some level, we understand that we are all connected to each other. Like our early ancestors, we know implicitly that we are a part of an interactive community and that our well-being and the community's well-being are connected. Perhaps this is not very different from what happens in plant and tree networks. As Wohlleben explains, when trees work together, they can create an ecosystem that is healthier by controlling temperature extremes and retaining moisture. They regulate their growth and the growth of others to jointly create a protected environment. Everyone benefits from sharing and working together. The whole is greater than the sum of the parts.

Studies of human cooperation, altruism, kindness, compassion, and empathy involve attempts to quantify and analyze subjective experience. They are often limited by the lens they are viewed through. Biologists look for genes or hormonal triggers. Sociologists and psychologists consider other behavioral markers, like reward or punishment. A different approach is offered by Nicholas Christakis. His recent book, *Blueprint: The Evolutionary Basis of a Good Society,* suggests that characteristics like kindness and cooperation and similar behaviors are universals. He calls this set of universals the "Social Suite" and says they are common in all human societies and are genetically encoded in humans and many nonhuman species. The Social Suite includes the following:

- the capacity to have and recognize individual identity
- love of partners and offspring
- friendship
- social networks
- cooperation
- preference for one's own group (that is, some degree of group bias)
- mild hierarchy (relative egalitarianism)
- social learning and teaching[20]

Christakis believes that because humans have always lived in social groups, essential social interaction with others shaped our genes. In other words, our genes have been shaped by the group and group dynamics. He says we are innately equipped to cooperate with others and are genetically disposed to be kind and reciprocal. Friendliness, he tells us, is hardwired into us. The genes that write the blueprint of our social life are a product of our evolution. Because of this blueprint, he says societies can only deviate to a degree from the Social Suite or they collapse.[21] And to prove his point, his book details numerous examples of human social experiments, both social group interactions that developed by necessity and those that were voluntary experiments. In other words, all successful human societies share the same characteristics.

Christakis believes that we share characteristics of the Social Suite with other social animals. Many other animals pair bond, have friendships, recognize individual identity, cooperate, and have some learning and teaching behaviors.[22] I suspect he would go so far as to suggest that certain Social Suite characteristics could also be applied to plants. While the characteristics of the Social Suite define our humanness in many ways, applying those characteristics to other species is not anthropomorphizing; it merely establishes that we share many essential social behaviors with other species. Those behaviors have a genetic component and have proven to have great survival value.

I like the way he defines friendship. He says friendship is volitional, long-term, and between unrelated individuals. It involves mutual affection and support—possibly asymmetric support in times of need. These relationships are characterized by trust, closeness, and affection. We take joy in our friend's well-being.[23] We know we are deeply connected. And this type of deep connection with others is an important aspect of our individual well-being.

He explains that friendship, identity, and cooperation support the capacity for social learning and teaching. Social learning is far more efficient than solitary learning. It is an important cooperative behavior, like other types of sharing behaviors. We benefit from the knowledge and memories of others.[24]

Western thought and the patriarchal religions that arose after the Neolithic agricultural revolution set humans apart from other animals and the rest of nature. Since the nineteenth century, there has been a movement both in science and philosophy to reintegrate humans into nature. Darwin argued that the forces of natural selection found in animals also apply to humans and their behavior. In contrast, Christakis argues that the universal Social Suite is also shaped by natural selection and defines our sources of happiness, including the social arrangements, that are good for humans. Christakis's thesis is that we are wired to be good members of a community. We are wired to share and be kind.

If we look around and all we see is endless fear, ignorance, and violence, we need to look again. Our media seldom report on the good news. But we all know that most people on a daily and hourly basis are putting their energy into doing good and serving others, including their families, children, friends, and, often, people they barely know or those who are complete strangers, as well as caring for other species and the environment. This is really who we are, but our media profits from emphasizing reports of the bad actors and dramatizing those who breach and exploit social norms.

I am not naively suggesting that violence is not a part of human nature. Psychopaths and sociopaths are far too common, making up around four percent of the human population.[25] These are individuals who lack a normal conscience; they have no empathy. They are pathological narcissists who do not care about others and whose actions are motivated by self-interest alone.

But exploitation and violence are not normal human behaviors. Christakis stresses that our underlying unity and the commonality of humans is based on love, friendship, cooperation, and learning. This is the fundamental good that lies in all of us. Humans, he says, have always had both competitive and cooperative impulses, violent and beneficial tendencies, but the arc of evolutionary history, he stresses, bends toward goodness.[26]

Research on trees has certainly demonstrated that trees have what appears to be friendships and familial bonds. They cooperate widely by sharing resources within communities and widely dispersed communication networks. Yet, it would be absurd to think that trees function as social networks in even remotely the same manner that we do. In fact, we know very little about these behaviors and their origin. There is no question, however, that a level of intelligent-like decision-making behaviors seems to be occurring in plant communities and forest networks. However, that does not mean they demonstrate intelligence as we understand it. Jeremy Narby, in his book *Intelligence in Nature,* suggests that "intelligence" might be too loaded a term to use when discussing behaviors of nonhumans, particularly for Westerners. He suggests synonyms like "smartness."

One important difference between us and our plant allies is that we have a centralized brain for processing information and plants do not. Stefano Mancuso, who is known as the world's top plant neurobiologist, points out that because plants, throughout their evolution, have remained anchored to the ground—that is, they do not move—plants have developed very differently. They have no centralized organs. That would be far too dangerous in a world filled with herbivores. Instead, their organs are distributed throughout the entire plant. He says they breathe with their whole body, they feel with their whole body, and they evaluate and make decisions with their whole body.[27] Animals, as he points out, merely run away from predators. Plants have evolved more sophisticated abilities to avoid dangers. Those behaviors involve the collection of extensive data, including chemical and physical parameters like light, gravity, minerals, moisture, temperature, mechanical stimuli, soil structure, gas composition, and other factors. They are amazingly sensitive. He believes that the entire root system of a plant acts like a collective brain. He suggests the term "distributed intelligence" as a way to understand how they receive and process information. A single grain plant may have millions of root apexes. An adult tree can have billions of agents exploring and gathering information.[28] Mancuso suggests that we might think of plant intelligence as similar to the group or swarm intelligence of social insects. He points out that for a long time we thought only animals like fish, insects, and birds demonstrated school-like, swarm-like, and flock-like behavior. It may be appropriate to consider whether those social behaviors could be applied to plants.

"Any group of single agents that makes autonomous decisions, lacks a centralized organization, uses simple rules to communicate, and finally, acts collectively, is similar to such a community," he states. That would include our green allies. But there may be more to plant intelligence than that.

If plant intelligence is collective rather than centralized, it functions as a network. Consider what we have learned about the transfer of information and resources through plants and their fungi partners and between the entire forest or plant community network: the wood-wide web. That may help us begin to grasp how network intelligence might work. But this type of collective intelligence is impossible for us to acknowledge if we cling to the anthropomorphic concept of what intelligence is.

Mancuso suggests we think of collective intelligence as a form of direct democracy. And direct democracy is not the same as representational democracy, which we know has limits and is subject to manipulation. He says that in nature, larger distributed organizations without control centers are more efficient that those with control centers. Decisions made by a group are better than those promoted and adopted by a few. So much for hierarchy and oligarchy.

Researchers Larissa Conrad and Tim Roper have studied group decision-making in animals and have found that group decisions are the norm and that the democratic method of decision-making is best for the community as a whole and produces fewer extreme decisions.[29] Bees, for example, decide where to relocate a hive by getting input from multiple scouts and then finding consensus. Conrad and Roper found that when animals share decision-making, they are pooling information. This offsets individual errors and increases accuracy. When those decisions involve conflicts of interest, they found that, contrary to intuition, conflicting goals increase both accuracy and the individual gains in ways that can improve outcomes for all stakeholders as long as they have some goals in common. The results are often better than conflict-free decision-making by animals who have the same goals. Diversity may lead to better decisions for all.

As humans, we may be moving toward a more collective form of intelligence. That was an initial promise of the internet, which has grown to be a massive non-hierarchical organization. Mancuso suggests that Wikipedia is a great example, with over forty million articles contributed by thousands of collaborators who find consensus collectively.[30] How very plantlike.

In humans, collective intelligence is understood to be shared, or group intelligence that comes from collective and collaborative decision-making. It can mean consensus or be reflected in cultural behaviors like voting systems, social media, and crowdsourcing. We now live in a world based on the internet. Media scholars Henry Jenkins and Pierre Levy say that collective intelligence is important for democratization because it interlinks with our knowledge-based culture sustained by collective idea sharing.[31] This contributes to a better understanding of the needs of a diverse society. Many researchers believe networks created by our information communication technologies enhance our social knowledge and often lead to better outcomes. One example is how people from all over the world are now collaborating online to address climate change through crowdsourcing. A few years ago, a PBS *NOVA* program featured examples of how millions around the world are now working together online to solve problems in a new, collaborative way. They cite examples from Wikipedia to open-source software to online citizen projects. The online game Foldit showed how in ten days, players from around the world helped produce an accurate model of a key protein found in an HIV-like virus, solving a problem that had stumped scientists for years. The MIT Center for Collective Intelligence is exploring the potential of crowdsourcing in addressing climate change through an online platform called Climate CoLab.[32] COVID-19 has shown us that collaborative problem-solving by scientists and communities is the new norm. The collective sharing of the COVID-19 DNA sequence and related research early in the pandemic led to breathtakingly rapid vaccine development in laboratories throughout the world. Putting competition aside, scientists came together for the common good, demonstrating what we are capable of when we join together and address complex problems. We may find that plant-like behaviors will help us solve many of our complex human-created problems. The use of holistic and synergetic approaches may be necessary to arrive at the complex solutions we need.

Christakis emphasizes that the Social Suite is founded on human evolutionary biology and suggests that it points to complex synergistic behaviors. He says it is both reductionist and holistic. He does not see humans as special but suggests that our closeness to animals reveals our common humanity.[33] He describes his system as reductionist in that it is based on biological principles and social facts, but it is also holistic for the role of emergence in our social systems and social lives.[34] When he speaks of the concept of emergence, he is referring to another stirring debate. Emergence tells us that wholes have properties that are not present in parts. An example he gives is the combination of carbon, hydrogen, sulfur, phosphorous, iron, and the other elements that create life. Life is a property not found in its constituent parts. And, of course, since it is not like a cake mix, we have not been able to duplicate it. Another example he gives is that consciousness emerges from a pattern of connections among the neurons of the brain. The brain itself does not create or give rise to consciousness. No neurobiologist has been able to create consciousness out of a mix of brain cells. Christakis suggests there is something else that happens in complex systems. He says a transcendent synergy exists among the parts in those complex systems.[35] That transcendent synergistic quality, which we will explore further in later chapters, may be why Christakis tells us that the evolutionary arc bends toward goodness. It may be why forest communities and plant networks by their nature cooperate, share resources, and appear to act altruistically. The ability to perceive holistically and act for the good of the whole may be an aspect of their intrinsic wiring. And it may be an aspect of our intrinsic wiring, as well.

That takes us back inevitably to the idea of consciousness and the varied debates about plant or "other" consciousness. New narratives are emerging. Scientists may be letting go of the antiquated idea that the brain creates consciousness, but they still cling to the idea that you have to have a brain to be conscious. So, let's instead consider intelligence. The question is, do plants have intelligence? Intelligence is not the product of one organ. Generally, intelligence is defined as the ability to acquire and apply knowledge and skills. It may involve skills like reasoning, perceiving relationships, and learning. With intelligence, we act after arriving at an understanding rather than acting based just on instinct. This implies "smartness." Plants do not have brains and neurons like we do. The vast majority of all species found on Earth, which are primarily plant species, perceive, learn, and react to new situations without a special organ called a brain. Animals like sponges have functioned for eons without a brain or nerve cells, and the recent popular fascination with the ability of slime mold to learn has caused some to rethink old concepts.

Plants have memory and the ability to store memory. They learn, and as Gagliano showed in her clever experiments, they can remember for as long as forty days. Her most well-known research involved *Mimosa pudica*. Mimosa is considered a sensitive plant; a threat will cause it to quickly close its leaves and play dead. This was thought to be a defensive "instinctual" reaction to reduce the effects of predators. In her experiments, she repeatedly dropped the plants twenty-five centimeters for sixty drops a session. The goal was to see if mimosa could learn from this experience. The plants were to perform a simple task: leaf folding in response to a perceived danger. Could they determine that dropping was an annoyance but not a danger? It turns out they could learn this. After four to six drops, they all stopped folding their leaves. Leaf folding takes energy, so they learned when it was necessary and when it was not. This was not mere instinct. And it turns out they also remembered for a very long time. Yet, they have no brain, so how is that possible?

Plants may know when to bloom and be able to determine the passage of time. One possibility is that they have epigenetic memory. Epigenetics refers to biological mechanisms that activate some genes in an organism while inhibiting others. In 2016, researchers at Los Alamos National Labs determined the sequence on RNA in plants that controlled the time of flowering in the spring. Cells may remember these epigenetic modifications. And these modifications, even though they do not involve alterations in the DNA, are inheritable.[36]

Plants have been found to be able to discriminate between positive and negative experiences and learn by registering memories of their past experiences.[37] Plants respond more slowly to stimuli than animals do, so it is very likely that we overlook a great deal of their sophisticated behavior. We just don't notice it.

Plant neurobiologists like Mancuso may be a bit out of the mainstream, but he is not entirely an outlier. Charles Darwin studied plants for most of his life and believed that roots act like a brain.

It is important to remember that Darwin was a botanist and naturalist as well as a zoologist. He wrote many books and dozens of essays about plants. He considered plants to be highly organized beings. He and his son Francis carried out hundreds of experiments on plants. They concluded plants respond to sensation and were, in fact, intelligent. In his book *The Power and Movement of Plants,* written near the end of his life, in his final paragraph, he summarized his results: "It is hardly an exaggeration to say that the tip of the radicle thus endowed (with sensitivity) and having the power of directing the movements of the adjoining parts, acts like the brain of one of the lower animals; the brain being seated within the anterior end of the body, receiving impressions from the sense-organs, and directing the several movements."[39] His son Francis provoked a storm of protest in the scientific community when, in 1908, long after his father's death, he repeated the claim that plants were intelligent.[40]

Much of the research on plant intelligence has been done on higher plants that have complex vascular systems, like flowering plants. But we have now identified brain-like entities in plant root tissues known as meristems. Current theories suggest that these tissues combine with vascular strands and are capable of electrical and molecular signaling similar to what is found in our nervous system. It is theorized that the plant hormone auxin, which is manufactured at the root and shoot junctures, may allow for the transfer of information through the plant. We know that in addition to auxin, plants produce many neurotransmitters and hormones, including dopamine, acetylcholine, glutamate, histamine, and glycine. Researchers have found that plants utilize many of the same neurotransmitters that we do. Higher plants have neurologically active compounds that play a role in their physiology and development. We know that in humans, neurotransmitters carry messages to cells and trigger responses. Neurological signaling molecules also seem to have a role in plants and plant communication. Some are produced in stems, leaves, nodes, cells, and roots in response to stress. Some are used to develop defenses against insects. Other neurologically active compounds that plants produce include caffeine, epinephrine, opiates, nicotine, and hyperforin, an antidepressant. In humans, the pineal gland produces the neurotransmitters serotonin and melatonin from the amino acid tryptophan. Many plants also produce melatonin and serotonin. Melatonin has been found in over 108 of the plant species used in Traditional Chinese Medicines and apparently has many functions in a plant's growth and development. Researchers believe that rapid cell-to-cell communication occurs in plants via bioelectric signaling and "neuronal circuits." It is interesting that our brain uses many of the same chemical compounds to affect growth and development that plants do. We are far from understanding these pathways, but it seems clear that plants communicate and transfer information internally in ways that, while decentralized, are not entirely dissimilar to the complexity found in animal brains.[41]

We know plants have senses, although they do not have organs like our eyes, ears, and noses. Mancuso explains that, in addition to our five senses, they have about fifteen others. They perceive a broader spectrum light. They intercept light, he says, and can determine its quantity and quality using molecules that act as photoreceptors.[42] Our light receptors are in our eyes. In plants they are everywhere: the leaves, the stems, the shoots, the tendrils.

They have a sense of smell but it is not concentrated in one area. The cells of their roots, leaves, and needles have receptors for volatile substances that set off chains of signals and communicate to the entire organism. Mancuso says we can think of plants as having millions of tiny noses dispersed throughout their structure.[43] We know that these smells are the words of a functional plant language.

Mancuso also tells us they have taste. They have chemical receptors they use to seek out minerals, salt, and other food substances in the soil. He says that their sensitivity to the taste of soil as they search for nutrients is far more developed than the sensitivity of our human palates. He explains that we know this because we have observed that plants produce many more roots where the concentration of certain mineral salts are the greatest. Plants use this sense of smell to penetrate to great soil depths to seek and find these nutrients.[44]

And yes, they may hear, too. Lots of animals, such as earthworms and snakes, do not have external ears; they feel vibrations in their jaws and in their heads as they slither on the ground.[45] The bodies of earthworms and other invertebrates can sense vibrations. They also detect light and tastes.[46] With plants, the vibrations in the earth and air are captured by their cells. These cells are sensitive to sound waves, which may indicate a type of hearing. And yes, they like certain music. Grapevines have been found to grow better, produce richer grapes, and require less pesticides when exposed to music. There is now a new branch of agriculture called "phonobiology." Grapes seem to like frequencies between 100 and 500 hertz. We also know that plant roots make clicking sounds, but not in response to the music—as far as we know.[47]

And, of course, they have a sense of touch. This has been documented by many researchers. The touch receptors are found throughout the plant. In many plants, like the widely studied *M. pudica,* it is a defense mechanism. Roots also have a sense of touch and respond to obstacles.

Plants may be far more sensitive to their environment than we are. Consider some of these extra senses that plants have. They can precisely measure soil humidity and find sources of water very far away—they have a built-in hygrometer. They can sense gravity. They can sense electromagnetic fields.[48]

We now know that plants are highly sensitive, with well-developed dispersed senses, including many we do not have. They respond to many types of stimuli. By most definitions, that makes them sensory beings. Arguably, they have a nervous system in that their phloem and xylem, which stretch through the entire plant, work like our circulatory system, and cleverly pump and transmits food, water, and chemical signals throughout the plant. They have brain-like functions that are dispersed throughout their roots—what Mancuso calls distributed intelligence. They do not have neurons, but they do have neuron-like receptors in their roots and elsewhere. These receptors respond to some of the same neurotransmitters found in our brains. That certainly implies sentience.

We are conditioned to think of plants as utilitarian objects, not as quasi-intelligent or sentient beings. As Mancuso explains, plants have evolved very differently from animals. They do not have individual organs like a brain, heart, lungs, and stomach. Just because they do not have a heart, we should not conclude they do not have a circulatory system. And just because they do not have a brain and spinal cord like ours does not mean they do not have a form of intelligence and some type of neural-like signaling and functioning.

The evolutionary path of plants was that of a stationary species that could not run away from predators. They needed dispersed organs so their brains, lungs, stomachs, and multiple sensory organs were not eaten by herbivores. They very cleverly developed distributed intelligence and dispersed organs that model and, in some ways, surpass our sensory paths. That is, as Mancusco suggests, true green brilliance.

Rethinking ingrained notions has consequences. If we accept that plants are sentient and have a type of intelligence, does that mean they should have rights? The idea makes most of us uncomfortable. Like most people, I eat both plants and animals. I also like to fish. When I catch a trout, I appreciate that a living being has died, and, as a result of their unwilling sacrifice, I have dinner. I try to balance any discomfort I may have by respecting their sacrifice and by holding a sense of gratitude. I am generally not as thoughtful about the vegetables I eat or the wood I burn.

Many indigenous cultures who were directly connected to the sources of food they hunted and gathered held a natural sense of respect and gratitude for all of life. Implicit in that was a sense of reciprocity. We now get our food from grocery stores and have lost any remnant of that deeper connection.

Robin Wall Kimmerer is a revered nature writer who is both a botanist and a member of the Citizen Pototwatomi Nation. In her book *Braiding Sweetgrass,* Kimmerer eloquently leads us to an understanding of what a reciprocal relationship with the plants that sustain us would look like. It is this depth of understanding that we may need to relearn to survive the climate crisis.

We are not there yet, but we are beginning to see the emergence of a sense of appreciation for both the animals and plants we share the planet with. The evolution of this deeper awareness began with animals in part because it is easier for us to relate to their pain than to any pain a plant may feel. Groups like People for the Ethical Treatment of Animals (PETA) have helped obtain more humane practices for the treatment of domestic animals. Yet they do not acknowledge that plants experience pain. I am not sure that there is adequate evidence that plants experience emotions and pain as animals do—concepts that have given rise to the idea of animal rights. Nevertheless, there has been some interesting movement in the direction of considering plant rights.

It is not a new idea. There are over four million Jains in the world. They refuse to eat any food that was obtained with what they feel is unnecessary cruelty. That includes root vegetables. They believe it is unnecessarily cruel to kill the plant by pulling out its roots.

In 2012, a river in New Zealand, including all the plants and organisms within its boundaries, was declared legally a person with standing to bring legal actions to protect its interests. The Swiss Government's Federal Ethics Committee on Non-Human Biotechnology concluded in 2009 that plants have rights and that we need to treat them appropriately. That means we humans have no right to alter their genes in ways that serve our needs. In the Netherlands, the single-issue Party for Plants entered candidates in recent elections.

I think the key is respecting plants and being grateful for what they give us. I like the way forester Peter Wohlleben addresses this issue. Gradually we have moved from treating animals as objects, as commodities. People are reducing meat consumption and promoting humane treatment of animals. He says parallels can be drawn. He asks if we can spare forest and plant communities unnecessary suffering. That would mean using forestry practices like selective cutting versus clear-cutting, applying species-appropriate management, leaving some old-growth forests, adopting organic agricultural practices, and taking a variety of other more holistic, sustainable agricultural approaches.[49]

If we do someday agree that plants are intelligent, a lot of revisioning will be required. We will certainly need to reconsider whether we are the only type of intelligence around. It may be useful to consider that there are other forms of intelligence that have been around longer than us, and it may be that they have something vital to teach us.

While much of this new information on what humans share in common with plants is fascinating, it is hard for us to fully grasp it or its implications. We are still at the cusp of a revolutionary new understanding. Plants truly are "other." We don't have to understand their unique senses or "other" intelligence. There are vast differences between us and the green species we share the planet with, and we must acknowledge that these green species provide us with our atmosphere, all our food, and most of our medicine.

We are learning that plants are far more than commodities provided for our use and benefit. Acknowledging that they may be sentient and have a form of intelligence may make us uncomfortable. The advances from new scientific frontiers always require some adjustment. Then our reality shifts. As we open our eyes to the implications of this new perspective, we may need to redefine words like "caring" and "respect." Learning to be comfortable with the meta-understandings brought by new perceptual breakthroughs often requires new stories as well. When the astronauts first reached the moon and sent back images of our beautiful blue and green planet, a new meta-understanding rocked our consciousness. We found new stories that shaped our understanding of ourselves and our place in the universe—stories that are still evolving. Now, we need new stories to help us develop more holistic and reciprocal behaviors—behaviors based on an understanding of interconnected relationships.

5 Greening Our Stories

Banyan (Ficus bengalensis)
The banyan. . . is one the wonders of the plant world. Its ability to form and send down aerial roots to make additional trunks to support the canopy enables a single tree to spread over several acres. . . . The banyan has been one of the most sacred trees of eastern Asia. The earliest Indian scriptures, the Vedas and Upanishads, link trees, and in particular the banyan, with Brahman itself—the immortal, innermost spirit of the universe. . . . The sacred scriptures themselves are regarded as the leaves of the Universal Tree. This makes it the Tree of Knowledge as well as the Tree of Life.

The Meaning of Trees

As we have seen there is an increasing scientific basis underlying our changing perceptions of the plant kingdom. First, it is now evident that established forests function as interactive, cooperative communities that communicate. And research suggests that plants appear to have attributes of intelligence, although their form of intelligence is very different than ours.

Most anthropologists believe that our hunter-gatherer ancestors, like many indigenous cultures, held animistic, nature-based beliefs. They saw all of nature, plants and animals, as alive—as sentient partners. They saw themselves as a part of the whole and ever-evolving fabric of life. As we have seen, those early beliefs lost their footing some ten thousand years ago, when our ancestors developed agricultural practices. They learned to domesticate both plants and animals. Once those plants and animals were under their control, early humans no longer gave equal status to their former allies. Their former allies became property and objects. Humans began to view other species abstractly, as objects of value that could be exploited, traded, and hoarded.

We have seen that after the cognitive revolution, our ancestors developed new capacities, including the ability to create shared fictions and mythologies. These mythologies were a unifying force, as large groups could now be bound together by their shared beliefs. These new beliefs placed their followers outside of nature. They were no longer part of a larger fabric of life. New gods emerged that gave humans a feeling of dominance over all other species.

These early beliefs gradually evolved into the religious systems now shared by most humans. Some of those religious systems are monotheistic. Many have creation stories that define the role of humans in relationship to the rest of nature, and many have dominant male deities, founders, and leaders. They have all developed into social systems and authoritative structures generally led by men.

Today, most people have a religious affiliation. The vast majority of us consider ourselves Christian, Islam, Hindu, or Buddhist. About 2.5 billion people consider themselves Christian, about half of whom are Catholic. About two billion identify with Islam. 1.2 billion are Hindu. 500 million are Buddhist. About 1.2 billion consider themselves nonsectarian, agnostic, or atheist. Millions of others identify with fifteen or so other traditions, including Judaism, Sikhism, and ethnic and traditional religions.[1] Traditional religions include Native American and aboriginal beliefs and folk religions.

We look to the religion we identify with for moral guidance. As a social species in a complex, diverse world, we seek ethical decision-making systems that reflect our values. Approximately eighty percent of the world's 7.7 billion people identify with one of the world's four major religions. All of these four major religions are slowly greening; that is, they are moving in the direction of a more unified view of our role within the natural world and our responsibility for stewardship, care, and protection of all of life. In some cases, this greener perspective has led to a reinterpretation of spiritual texts, including creation stories, as well as urgent guidance to both individuals and policy makers as the potential impact of the climate crisis is becoming more apparent.

These four religions all have vast and varied traditions, with many sects, denominations, and variations. There are over 200 separate Christian denominations in the United States alone. All Christian traditions share certain sacred texts and scriptures encoded in the Bible. They also share the creation story found in Genesis. This creation story has long been interpreted by some practitioners and religious scholars as a mandate for making all of nature subservient to man. This mandate, they believe, directed us to subdue nature and gave us dominion over all beings. But does it really say that?

I am unequipped to venture into the world of biblical interpretation. Suffice it to say, with the exception of certain fundamentalist and evangelical branches of Christianity, "dominion" is generally now understood to mean stewardship and responsibility—terms that dramatically change the story but that may be closer to what was originally expressed by the ancient Hebrews who gave birth to the book of Genesis. Scholars like Walter Brueggemann say that early Hebrews had a three-part covenant: to God, to humans, and to the land. Humans were tasked with caring for and cultivating the "garden." They were to watch over and preserve all of the community of creation as God's agents. That is what was originally meant by "dominion."[2]

In 2015, Pope Francis published his second encyclical, *Laudato si'*, subtitled, "On Care for Our Common Home." A papal encyclical is considered a clarification and amplification of a viewpoint and is intended as guidance to Catholics and the global community. In this eighty-page letter, the Pope criticizes our excessive consumerism and the irresponsible overdevelopment that has led to our global crisis. It is a call to arms in which he calls upon the global community to take "swift and unified global action."

He bases much of his letter on his namesake, Saint Francis, and on the church's interpretation of Genesis. He carefully lays the framework for this interpretation, calling on the words of his predecessors as well as others, such as contemporaries like the Patriarch Bartholomew of the Orthodox Church. He says that the symbolic creation narrative in Genesis is a profound teaching describing the interconnected harmonious relationship between the Earth, our neighbors, and God. He says that by distorting the mandate and the meaning of "dominion," we have disrupted the harmony of creation, the harmony that Saint Francis experienced. He states that there is no justification, biblical or otherwise, for our exploitation of nature, our domination over other creatures, and, by implication, our anthropomorphism.[3]

He says our job is to "keep and till" the garden of the world and stresses that keeping means "caring, overseeing, protecting, and preserving." He states that this "implies a relationship of mutual responsibility between human beings and nature. Each community can take from the bounty of the earth whatever it needs for subsistence, but it has the duty to protect the earth, and ensure its fruitfulness for coming generations."[4] Because human beings are endowed with intelligence, he tells us, "we must respect the laws of nature and the delicate equilibria that exist between the creatures of this world."[5] As the Pope explains, the language of Genesis tells us that the proper role of humans must be viewed in relation to the planet as a whole and all of life. If a Catholic practitioner were to deviate from this guidance, they would be committing a sin, he implies. So, what do these words really mean? They could be subject to more than one interpretation. That is why the Pope's detailed eighty-page teaching, filled with examples, is exceedingly helpful.

First, he astutely identifies multiple aspects of the climate crisis and stresses that we must pay attention to what science is telling us. The crisis is human-caused, and he minces no words. He states that these causes include the explosion of technology without ethical guidelines, sustainable standards, or self-restraint,[6] and blames the perception that reality can be limitlessly manipulated, reductionism, the subject–object dichotomy (what he calls an undifferentiated one-dimensional paradigm),[7] practical relativism (seeing everything through the lens of self-interest), and misguided anthropocentrism (dominion, not stewardship).[8]

He introduces several new paradigms. One is "integral ecology," which is considered a new paradigm of justice, both in how we treat the environment and how we treat other humans. He says that nature is not separate from us. Everything is connected. Fragmented and isolated knowledge must therefore be integrated into a broader vision that considers the interrelation between ecosystems and the spheres of social interaction.[9] We are encouraged to embrace integral ecology as a practice for everyday life. He says it is inseparable from the principle of the common good both with respect to the environment, with regard to the poor, and with regard to future generations. This, he implies, is what social justice requires.

In the final chapters, the Pope says that we must reshape our behaviors and habits. This involves a new lifestyle in which we overcome collective selfishness and obsessive consumerism, which he says arises from an empty heart. He calls for an "ecological conversion" and an ecological spirituality that will motivate our passion and concern for the protection of the world. He says purchasing (and all consumption) should be viewed as a moral act; that we must make a habit of assessing the impact of our actions and decisions on the world around us. This conversion, he says, must be inspired by an inner impulse that motivates and gives meaning to individual and communal activities.[10] Conversion involves recognition of our errors, repenting, and changing to attitudes that foster gratitude, generosity, and loving awareness of the connection with all creatures and the virtues of moderation, simplicity, and humility.[11]

With finality, the profound teaching encapsulated in *Laudato si'* throws out the old, distorted creation story and beautifully reframes it for the twenty-first century. There is now a new and powerful story. There should be no more confusion. We are a part of nature and have a unique responsibility to protect and care for all of life. This slow greening is contagious and spreading around the world. But, of course, there is far more in the Pope's letter. It is highly recommended to all.

But there is another aspect of Christianity that needs to be addressed: the concept of redemption. As the late renowned theologian and evolutionary historian Thomas Berry explains, redemption tells us that we are not of this world but that we belong to some transcendent world. We are redeemed from both sin and nature. It says that the natural world is not sacred; it is an objective reality that we are to view as subservient to the transcendent, high spiritual reality.[12] Variations of the concept of redemption are found in many major religions. They can be viewed as a hindrance or detriment to much of the greening occurring in the landscape of world religions.

The World Council of Churches is a fellowship of hundreds of Christian churches, with over a half a billion members. Far from transcendence, their work involves what they call eco-justice, addressing the connection between justice, poverty, and ecology in the here and now. They are very involved in climate change issues and building sustainability. If there were ever an eleventh commandment, as some have suggested, it would be embodied in their principles.

As a former Catholic, I am thrilled by these teachings, which show a new greener Christianity, particularly the *Laudato si'*. I wish I had heard terms like eco-justice and integral ecology from the pulpit in my youth when I struggled with making sense of the world and my place in it. As a young environmentalist in the early 1970s, my questing heart and mind found only a painful dissociation and irrelevance in the catechism I had been taught. Prompted by astute critiques like Lynn White's now classic "The Historic Roots of Our Ecological Crisis," published in 1967, which linked the Western world's exploitive attitude toward nature to medieval Judeo-Christian values based on dominion over the Earth, viewing nature as having no value apart from what it provides us, and giving us freedom to exploit, I realized I had to, in good consciousness, look elsewhere. I began a long spiritual journey though theosophy, anthroposophy, shamanism, Hinduism, and Buddhism. Somehow, I missed Islam.

Islam is the second-largest world religion and is the state religion of twenty-seven countries. Professor Odeh Al-Jayyousi, who is head of innovation at the Arabian Gulf University in Bahrain and a member of the UN Global Science Panel, says that the Islamic worldview offers a model for a transition to sustainable development because it focuses on justice, de-growth, and harmony between humans and nature. De-growth is a political, economic, and social movement based on ecological economics, anti-consumerism, and anticapitalistic ideas. We know that the environmental crisis is largely a moral and ethical crisis. Al-Jayyousi says that the Islamic worldview defines the good life as living lightly on the Earth and caring for both people and nature. He believes that the Earth will find balance only if we rethink our lifestyles and mindsets.[13]

Iranian-born Seyyed Hossein Nast, professor of Islamic Studies at George Washington University, says that environmental ideas have gained traction in the Muslim world, particularly in Turkey, Iran, Indonesia, and Malaysia, and that there is a lot of activism in Pakistan, Morocco, Egypt, and Nigeria. He says that contemporary green Muslims believe the teachings of the Prophet Muhammad have clear implications for and applications to our current crisis. Major Islamic texts are filled with references to nature and its sacredness. He says that the Prophet encouraged thoughtful stewardship, instructed on water conservation, spoke about avoiding wasteful consumption of resources, advocated for proper land use, promoted the need for compassion for all animals, and made many specific references to the stewardship of trees.[14]

"Do you not observe that God sends down rain from the sky, so that in the morning the Earth becomes more green?" This passage from the Qur'an (Sutra 22:63) is a beautiful metaphor. According to Frederick Denny, Professor of Islamic Studies at the University of Colorado, it tells us "the color green is the most blessed of all colors for Muslims and, together with a profound sense of the value of nature as God's perfect and most fruitful plan, provides a charter for the green movement that is the greatest exertion yet known in Islamic history, 'a green jihad' appropriate for addressing the global environmental crisis."[15] She says that God entrusted humanity with the duty to protect and restore balance in the environment and to protect it for future generations.

The Qur'an may offer a way to a sustainable world, but that viewpoint is not always emphasized. In many Islamic countries, there is a divergence between what green Muslims understand from the teachings of the Prophet and what the clerics are ordered to preach. Islam is fairly diffuse and is rooted in many developing countries, including some that have fossil-fuel economies. It has not been easy to unite Muslim countries around climate change, but many groups worldwide are involved with working on this greening.[16]

The third largest religion, Hinduism, is getting greener too. That is no surprise to me. My own experience is that Hindu teachings are alive, colorful, and embodied. Traditional Hinduism sees everything as alive. Every living thing has an *atman,* or soul. Everything has consciousness. It is a religion with many different forms and traditions but no governing body, holy book, or central order. This may seem chaotic to our logical minds, but that is part of its allure and effectiveness in helping dissolve the constraints of the ingrained Western intellect.

Hindu beliefs are similar to those of Buddhism, which are discussed below and include the idea of dharma (ethics and duties), samsara (the continuing cycle of birth, life, death, and rebirth), karma (action, intent, and consequences), and moksha (liberation from samsara), as well as various practices or yogas. These practices, Hindus say, lead to union with pure or cosmic consciousness.

To Hindus, everything has a soul or true self, and that true self is eternal. The goal of life, according to some of the nondual practices I am most familiar with, is to realize that there is no difference between your soul, or atman, and the universal atman (Brahman). Nondualism tells us there is no separation; that is, the dichotomy between self and other is transcended. It is a state of awareness referred to as "one undivided without a second." There is just one consciousness, or supreme soul, and it is present in everything and everyone. That means that all life is interconnected. Everything is in a state of oneness. Everything is divine: humans, trees, rivers, animals, the rising sun, friends, family. Because it is all divine, everything is sacred and worthy of reverence. There is no competition between man and nature but a unifying divinity that connects everyone and everything. Most Hindus are vegetarian. Traditionists practice *ahimsa* (nonviolence). The essence of this teaching is respect for all life. They believe divinity is in all things, including plants and animals. This all seems in stark contradiction to the current ruling party's push for Hindu nationalism and their harsh treatment of their Muslim minorities.

Hindus do not feel they have authority or dominion over nature. But they do believe they are subject to a higher and more authoritative responsibility for creation. A key aspect of this is the doctrine of ahimsa. The Vedas are the early religious texts that originated in India about 1500 BCE and are the basis of Hindu teachings. Among the Rig Veda teachings is the concept that trees and plants possess divine healing properties. Many Hindus believe that every tree has a Vriksha Devata—a tree deity or spirit. They worship the tree deities with prayers and offerings. They do not view them as gods but as manifestations of the divine. Tree-planting is considered a religious duty.[17]

India has more Hindus than any country. I traveled the subcontinent in the era before cell phones and the country's movement to modernize. I was on a quest, like so many Westerners, seeking the wisdom of living spiritual masters and adepts. From small temple towns cut off from the rest of the world to large ashrams milling with the masses, I pondered. Gradually, my rational mind was slowed and balanced. There was something inexplicable and vibrant in the land itself. It was reflected in the genuine, uncontrived nature of the many natives I encountered who showed me basic kindness and a sense of connection. From the sadhus, their eyes in ecstasy, to ordinary householders filled with grace, I felt I was in the midst of a living religion—that is, until I considered the environment around me. It was simply trashed and trampled. Litter was everywhere, along with open sewers and horrid air quality. My lungs barely survived several visits to New Delhi, where then the air quality was so bad you could not see the ground from 20 stories up. Flying into San Francisco Bay on my return and seeing its emerald-green pristine hills and valleys and the brilliant clean blue bay, I burst into tears out of sheer relief.

India and the Hindu world are now slowly greening. NASA images show that much of India is literally turning green as a result of tree-planting programs and better agricultural practices. In 2009, the first Hindu Declaration on Climate Change was adopted by spiritual leaders. In 2015 at the Paris Climate Conference, the Hindu Declaration on Climate Change declared: "We call on all Hindus to expand our conception of Dharma. We must consider the effects of our actions not just on ourselves, and those of humans around us, but also on all beings. We have a dharmic duty for each of us to do our part in ensuring that we have a functioning, abundant and bountiful planet."[18] As Makarand

Paranjape of the Nehru University in New Dehli says, "life on Earth challenges us to transform ourselves personally, socially and ecologically." But, he says, many Hindus have lost their way and need to learn how to be Hindu again in a way that is ecologically responsible.[19]

There are also new stories arising in Buddhism. Buddhism does not have a creation story or creator God, nor does it have a central leader like the Pope, and it contains many traditions and beliefs. But in recent years, individual western Buddhist teachers and a few writers and scholars have begun to interpret Buddhist teachings in a greener manner.

Many Buddhist teachers refer to the metaphor of Indra's Net, a net that is made entirely of brilliant jewels that all reflect each other. Each jewel is reflected in the whole. This is a metaphor for a view of the cosmos, where the totality is a vast body of members, each sustaining and defining each other. This is an apt metaphor for the emergent greener Buddhism.[20]

Traditionally, Buddhism says our problems, including the environmental crisis, are due to our karma, ignorance, and delusions that lead to cravings and aversions and ultimately to suffering. We are controlled by our habits and unconscious patterns, which act as filters through which we perceive reality. And since the filter is not clear, what we perceive is false. These teachings help me see the degree to which my beliefs, what they call "view," shape my conduct. As a member of an elite material society, constantly bombarded by media telling me I need things to be happy, it becomes difficult to separate fact from reality. I often become unconscious and simply react.

I have struggled as I have explored Buddhist teachings. I am still a beginner seeking to understand these complex, abstract teachings and practices. I am in love with the world. I have no wish to transcend it, and I do not merely see it as a source of suffering. So, I have been heartened by the greening that is happening within Buddhism. A new concept is emerging, called ecodharma, and it is shaping a new story.

"Dharma" means teaching, and like in Hinduism, it is the basis of ethics and morals. Ecodharma explores Buddhist teachings in a way that helps practitioners deal with the climate crisis and social concerns. In some ways, ecodharma goes back to the teachings of Zen poet Gary Snyder. He taught that we need a proper understanding of where we are in the Big Picture, the Earth mandala in which we stand, so we are motivated to live more harmoniously. Joanna Macy, climate activist and proponent of deep ecology, which embraces the inherent value of all living beings, talks about the ecological self as relational and interdependent, in interaction with all beings.[21] This ecological self is one node in a web of relationship, like Indra's Net. Many are familiar with Zen Buddhist Thich Nhat Hahn's teaching on "interbeing," the implicit interconnection of all things. He also emphasized the need for "engaged Buddhism," the idea of taking the insights that come from the practice of meditation and dharma teachings into the world to address social, political, and environmental injustice. Buddhist scholar Kenneth Kraft uses the term "eco-karma" to describe the impact of human choices and their effects. Stephanie Kaza, author of *Green Buddhism,* discusses the many ways Buddhist teachers and practices are greening in response to the realities of the twenty-first century. She points out that some environmentalists and Buddhists now consider themselves "ecosattvas," or environmental warriors, those who have taken up the green path of service to all beings in response to the planetary crisis. The ecosattva concept comes from the Mahayana Buddhism term "bodhisattva." Mayahana Buddhists believe in reincarnation, and a bodhisattva is an individual who has reached enlightenment but who vows to continue to reincarnate to help all beings who are suffering, even if it takes an inconceivable number of lifetimes. Ecosattvas reinterpret that vow as a pledge to serve the interests of ecological balance and harmony, no matter how long it takes.[22]

For me, Buddhism is a deliberate process of mind-training and mind-taming. Some consider it a practice rather than a religion. Therefore, it is a tool we can all use to develop greener behaviors. Meditation helps us still and calm the mind so we begin to deconstruct our negative patterns and behaviors and some of the delusionary and destructive ways we view the world. One of those delusions is that we have a separate self. This self is an inner mental construct that allows us to objectify the outer world. It seems real but it is not. It is merely a collection of behaviors, likes, and dislikes. It causes us to identify with objects outside ourselves that we crave.

Greening does not come easy to Buddhism. David Loy, Zen teacher and author of *Ecodharma,* suggests that cosmological dualism (the dual states of samsara and nirvana) and the individual pursuit of enlightenment that are common in classical Buddhism and modern mindfulness training both encourage indifference to social and ecological problems, rather than addressing the problems of the world and helping it transform.[23]

Loy tells us this constant grasping never fills us but causes us to further devalue the natural world. Buddhist practices help us understand that this separate self is not real, so we can shift our focus to the real task of making the world better.[24] It is hard work and is, at times, grueling to see and dissolve our many illusions, but meditation and contemplative practices do help us cut through our personal delusions and soften our sense of self and self-importance.

Likewise, Loy tells us, these delusions are collective as well as individual. They create a sense of duality between humans and the rest of the biosphere, leading to unsustainable, corporate consumer culture and a collective sense of alienation and separation from the natural world. We need to cut though these mental constructs and embrace the actuality of our nonseparation (nonduality). Being human, he suggests, means that we are part of something greater than ourselves and that our role is to serve the well-being of that whole—the path that will also heal us.[25]

That sounds remarkably similar to what Pope Francis says and what Islam and Hinduism also teach. So, it is the new Buddhist story too.

The only solution, as Loy tells us, is to recognize that we are an integral part of the Earth. He says we need to understand that "we never left nature . . . we are nature. The Earth is not only our home, it is our Mother. It is the Source."[26] He implies Buddhists need to stop trying to transcend. There is nowhere to go but the here and now. That changes the story.

In other words, the next step Loy and other Buddhist teachers suggest is a collective awakening where we awake to the understanding of the larger whole, of our nonseparation with Mother Earth; that is, we realize we are part of the whole, of the fabric of all life. That this is true nonduality. There is no separate pure consciousness—it is all one unified living fabric. Nirvana is right here, right now. Loy says many Buddhists may have been thinking of enlightenment too narrowly and looking for it in the wrong places.[27]

And what about the trees and plant communities that make our existence possible? How are they reflected in this greening of our stories? Pope Francis refers us to the revelations of Saint Francis. The Canticle of the Sun is a song that Saint Francis composed in the ecstatic state he was habitually in. This Christian mystic, beloved by Christians and non-Christians around the world, wrote this canticle in 1224. It is considered to be among the first works of literature. He was blind at the time, so it was dictated to a companion. Although he could not see, his mind rested fully in the presence of inner light and awareness known in all spiritual traditions as pure consciousness.[28]

In the Canticle, Saint Francis praises God through all creation as he sees the divine everywhere in the natural world. The divine is Brother Sun who brings the light, is Sister Moon and the stars, is Brother Water, is Sister Wind, is Brother Fire. The divine is Sister Mother Earth, who sustains us and governs us. She governs us—it is not the other way around! Saint Francis was a nature mystic who taught love and empathy for all of creation, from God to soil and rocks. He not only cultivated plants but also cared for wild plants and grasses and insisted they have a protected space.

He had what is known as a second spiritual conversion around 1213. Thereafter, he also preached to both plants and animals. He preached to flowers, cornfields, and vineyards. He spoke to plants as if they were endowed with reason, and he said nature spoke back to him. He lived the conviction of cosmic kinship, holding that all living things were brothers and sisters because they have and were part of God's family.[29]

Buddha, according to tradition, became enlightened under the Bodhi tree, a type of fig. As the story goes, in the course of his meditations, Buddha was tempted by the demon Mara (a representation of one's ego) and was challenged by her to defend his claim of enlightenment. When she asked who bore witness to his achievement, Buddha placed his hand on the Earth; he touched the Earth under the Bodhi tree. This symbolized that he realized that the Earth was a fully conscious living being and able to recognize and affirm the level of his consciousness, his enlightenment. Buddha saw all of the Earth as conscious. The Bodhi tree in Buddhism symbolizes wisdom, compassion, and awareness.

The concept of the tree of life is a widespread myth or archetype found in many world religions. In Christianity's Garden of Eden, it is the source of eternal life, not to be confused with the tree of good and evil. It is also a symbol of Christ. In Islam, it is the tree of immortality or ownership. In the Islam story of the Garden of Eden, the tree was forbidden, as the idea of ownership was forbidden.

The tree of life is associated with wisdom and consciousness. Among the pre-Columbian Mesoamerican cultures, the tree of life was the world tree, the axis mundi, that connected all the cardinal directions, the upper world, the terrestrial world, and the underworld. It represented their entire cosmology and the path to consciousness.

It is more than a metaphor or archetype. Years ago, while studying different shamanic traditions, I visited some of the caves in the Yucatan. Our group spent a night in prayer and ritual working and walking deep underground with traditional teachers. At daybreak, we arrived at the most extraordinary vision. We were in an enormous cavern, facing a massive underground tree, whose roots extended into the unfathomable depths and whose branches reached up toward the small rays of light permeating down on us from some seemingly cosmic source. The source of this amazing light was a small sink hole at the surface that quite magically sent down light rays to this extraordinary world tree. This tree of life, found deep underground, made of stalactites and stalagmites, was real, radiant, and transformative. I understood both the magic and the metaphor, but I can only imagine what it was like for the ancient Maya to spend the night as we had, walking and praying in this massive cavern and at dawn seeing the vision of the light-filled tree of life.

Among Native Americans, the tree of life was a path to the spirit world. Hindus continue to consider trees divine.[30] In Jewish mysticism, the tree of life forms the central symbol of the Kabbalah, representing the powers of the divine realm.

While early Buddhism determined that trees had neither mind nor feeling and could be lawfully cut, they hedged their bets and recognized that trees may have spirits who reside in them. Hindus continue to believe in tree spirits and take care to perform proper rituals before a tree is cut, and certain holy trees are not to be cut at all. Christianity and Islam considered the worship of trees idolatry. Various cultures like the Druidic and Celtic cultures valued trees as spiritual beings. In my travels, I visited some of the old Druidic sites, including Stonehenge. I found it interesting that while Stonehenge was spectacular, it was the nearby grove of trees, which also seemed quite ancient, that drew and held my attention. Some ancients considered trees to be oracles who spoke to them. In earlier days, religions other than the main emerging four were labeled as pagan or worse.[31]

That is the overstory. All of these stories tell us there is a deep intimate connection between humans, trees, flowers, and all of the plant world. It is an ancient connection deep within our consciousness. The understory is more expansive but hidden, like the story of what happens below the canopy of trees. It is rich, deep, and communicates with the deeper unconscious parts of our psyches. The overstory is that new belief systems spread throughout human culture several thousand years ago. In time, they took the form of our major religions. But below the canopy, the older beliefs they replaced were not entirely lost. They were only disguised, covered over by a new veneer. This has been well documented time and again. Our species, *sapiens,* has been around for about 177,000 years. For all but a tiny fraction of that time, we lived in small groups and in nature and were intimately connected to the plants and animals in our ecological niches. We may not have known about photosynthesis, but we knew which plants to eat and which to avoid. We knew which provided medicines. We knew when and how to hunt and when and how to avoid being hunted. For these early humans, nature was not just a source of life; it was also a source of awe and amazement. There were special places of reverence that our ancestors felt held unusual powers or energy. Those places of power might be natural formations, groves of trees, or other areas where they gathered and worshiped. It is well documented that Christian churches and cathedrals were intentionally built on the sites of much older temples and power points. The Chartres Cathedral is one example. Older beliefs were also embedded into the traditions of the newer religions. Christmas overlaid the solstice. That is why we bring a tree into our homes in December. Easter overlaid the equinox, and we color Easter eggs and recognize the fecundity of the Easter Bunny to celebrate the fertility rites of spring. After the conquistadors overtook Mesoamerica, the Guadalupe figure overlaid an older Aztec goddess—a common practice. The Q'ero, descendants of the Peruvian Inca, continue to quietly perform their traditional rituals, often in Catholic churches. In Tibet, Buddhism still incorporates many traditions from Bon, an older Earth-based religion.

There is one more understory—one that is buried even deeper in the roots of our religious prehistory. Our early ancestors long ago discovered that certain plants and fungi could induce ecstatic states. They found a set of plants now called entheogens, which means "the god within," that were able to induce just such a state. These special allies opened the doorways of imagination and inner journeying in our ancestors. The entheogenic theory suggests this is how religion developed.

The term "soma" is found in some of our oldest religious texts, the Vedas, which were compiled in 1500 BCE but believed to be much older. The Avesta, a Zoroastrian text from around the same time period, mentions "haoma." Both substances were derived from plants. A drink of soma was said to be consumed by the gods, giving them fantastic powers: "We have drunk the Soma and became immortal; we have attained the light."[32]

Ethnographer Gordon Wasson proposed that soma was derived from psychoactive mushrooms. Others suggest cannabinoids or a mix of alcohol and other natural substances may have induced religious experiences.

And then there were the Eleusinian Mysteries, the rituals, ceremonies, and experiences of the Greco-Roman period celebrating Demeter, the Mother of God. They were intended to elevate man to the divine sphere and, apparently, they did. They were practiced for some 2,000 years. These mysteries involved fasting and drinking a drink made from barley and pennyroyal, but not just any barley. Ergot is a parasite that grows on barley, creating a psychedelic similar to LSD, and was thought to be the source of the secret aspects of the mystery where many had powerful visions, ecstatic states, and experienced a transcendent reality.[33] Participants came from all over the world and included historic figures like Sophocles, Aristides, Cicero, and many others who would shape the emergence of our world culture and ultimately its religions.

It is unclear whether religions were the opiate of the people (as Marx stated), or whether it was the other way around—that the opiates and their psychedelic plants and their fungi sisters were the religion of the people. Regardless of what overstory or understory we identify with, our religious beliefs were shaped by nature, and in particular shaped our understanding of our place within the cosmic whole.

To some degree, whether we acknowledge it or not, we are all a bit like Saint Francis. We have all been deeply influenced by nature. Our connection with nature may be genetically encoded. Biophilia may be in our genes. Some of us are absolute nature mystics. Many of us have forgotten who we are and where we came from, but love of the world we are a part of is still in our essential nature. We will awaken—we will remember. All the major religions are now trying to help us.

All religious paths lead to consciousness. And you certainly do not have to ascribe to a religion to seek and find consciousness. Our major religions and the revelations of science (which we discuss in the next chapter) are now telling us a greener story. That story says we are innately programmed for consciousness—a consciousness that recognizes we are connected with all of life, that we are not separate from nature. We are, therefore, innately all biophiliacs.

6
Seeing with a Greener, More Humble Lens

Cedar (Cedrus)
The most ancient cedars of Lebanon may be 1000 years of age. . . . [The Epic of Gilgamesh] relates how a man named Gilgamesh visits a cedar forest . . . and, falling prey to his own greed and desire for fame and eternal life, destroys the forest and its spirit guardian, the giant Humbaba. But the consequences are dire, for he finds himself also responsible for the death of his dear companion, Enkidu, and in bitterness and despair loses his own life as well as his afterlife. The ecological moral in this legend is obvious.

The Meaning of Trees

While we may be wired for biophilia, some of us seem to have forgotten or lost our green lens. I was shocked to find that we humans have yet another deficit. Many have developed what is known as "plant blindness." You heard me right. A significant number of people have a cognitive bias that makes them tune out the green world: the plants and trees around us. To them, the green world that supports us is, at best, background noise. This happens early. Children are far more interested in animals than they are in plants. They can actually lose the ability to notice plants in their environments. The term "plant blindness" was coined in 1998 by Elisabeth Schussler and James Wandersee, two botanists and biology educators.[1]

Many humans have this bias. Not only do they not notice plants in their environment, they also do not recognize that trees and all plants are critically important to all of life. They think of the plant kingdom as inferior or irrelevant. It is an aspect of what is clinically known as nature deficit disorder. People who suffer from this deficit find it hard to even detect images of plants. They unconsciously filter out our green allies. While the idea of plant blindness may sound like a metaphor, it is not. It is an actual syndrome.

We know that this is partly the result of increased urbanization and because modern humans spend so much time looking at little handheld devices that they consider important to their well-being. Many people, according to BBC journalist Christine Ro, have become increasingly alienated from nature.

Most nature programs are about animals. Young children love cuddly little stuffed animals. Adults also seem to be zoo-centric; they tend to like animals and ignore plants. Students are no longer signing up for plant biology classes, and public funding for plant science is drying up.[2] Plants make up fifty-seven percent of the federal endangered species list but only get four percent of the funding.[3]

We don't really know why this disconnect happened. One theory is that since plants often grow close to each other, we visually blend them together.[4] And they don't move—so that could be a factor. It could also be a learned behavior. Environmental professor Kathryn Williams points out that because textbooks give a lot more space to animals, some students may conclude that plants do not matter that much.[5]

Some researchers actually think that we may be wired for plant blindness. Humans are more supportive of the conservation efforts for species with human-like characteristics. According to Schussler and Wandersee, the brain is a difference detector.[6] Plants are not different enough. Plants are not threatening, and we tend to group them together. Large animals with big eyes that are more similar to us, like polar bears, are found to be more interesting. Because they are similar to us, their plight touches our hearts, evoking empathy. It is harder for many people to connect emotionally with plants.[7]

She suggests several strategies for developing more plant empathy. Since they don't have faces or move around, we need to consider how they are similar to us. Art, imagination, and ritual can help us emotionally connect with plants. We now know that plants are communicative; they have more than fifteen sensory modalities; they send food to their young, the elderly, and the infirm; they learn; they have friendships; and they feed and maintain entire complex communities. We do not have to anthropomorphize to make them seem like us. They are like us in many ways. And we have evolved to be plantlike in many critical ways. So, how can we see green again?

I would suggest that we need a much better grasp of how critically dependent human life—indeed all planetary life—is on the green world. We need to start seeing through a greener lens. Recall that every day in summer, trees release twenty-nine tons of oxygen per square mile in an average forest.[8] Two mature trees can provide enough oxygen for a family of four. Of course, not all trees produce the same amount of oxygen in the same amount of time.[9]

Let's say you were trapped in a ten-by-ten-foot chamber. How much oxygen would you need? And how would you get rid of the toxic carbon dioxide you emit? You would need about 300 to 500 plants to produce the right amount of oxygen, but it's much harder to estimate the amount of carbon dioxide the plants absorb, especially if every time a person breathes out, they inhibit oxygen production. To be safe, you would need about 700 potted plants. That ten-by-ten space will be very full.[10]

Trees also filter out tons of particulates and toxic chemicals from the air. According to Ecosia, an organization that plants trees all over the world, our forests absorb about a third of all global emissions every year. They explain that particles, odors, and pollutant gases such as nitrogen oxides, ammonia, and sulfur dioxide settle on the leaves of a tree and that those leaves absorb these toxic chemicals through their stomata, or pores, effectively filtering these chemicals from the air.[11]

We are pumping an enormous amount of carbon dioxide into the air, which in turn is causing global temperatures to increase dangerously. Researchers have found that about twenty-five percent of the carbon emissions we generate are absorbed by our green friends. The oceans also absorb an enormous amount.

A 2014 study by Ying Sun, et al., published in the US journal *Proceedings of the National Academy of Sciences,*[12] suggests plants could take up sixteen percent more carbon than we thought and, therefore, that some climate models underestimate how much carbon is stored by plants. As a consequence, people may overestimate how much carbon goes into the atmosphere. The ability of nature to take in and retain carbon from the atmosphere is called the carbon sink—and the carbon sink may be bigger than we thought. The difference is due to the observation that CO_2 moves around in the plant's leaves.

These results were summarized in the online journal *The Conversation.*[13] We know that as the rate of CO_2 in the atmosphere increases, photosynthesis also increases, to a degree. But after a certain point, the plant is saturated with CO_2. The study focuses on how CO_2 moves through leaves. It actually has to move through several membranes to get to the chloroplasts, and this movement slows the CO_2 saturation point and may result in a sixteen percent increase in the carbon land sink.

The researchers also found that about fifty percent of all CO_2 taken in by photosynthesis goes back to the atmosphere soon after, through plant respiration. Of what remains, more than ninety percent eventually returns to the atmosphere through microbial decomposition in the soils and disturbances such as fires. But what stays remains in the carbon land sink, and that amount is more than we thought.

Plants make plant sugar from sunlight, and the rest of us are entirely dependent on them. Almost everything on Earth either eats plants or eats animals that eat plants. That includes us. No plants, no food, no humans. Even fungi depend on plants for their food. There are a few bacteria and single-celled animals that also photosynthesize, but they don't add much to our food chain. All of our nutritional needs are met due to our green allies.

Without trees, most of the planet would be a desert. When ocean water evaporates and forms clouds in the atmosphere, almost all of it falls as rain within the first 150 miles of a coast. The fact that water in the form of rain reaches massive inland areas is entirely due to coastal forests. When rain falls, plants absorb the moisture and hold it in their bodies. This moisture is more than they need to photosynthesize, so some is lost through transpiration and evaporation. This lost water fills the clouds with moisture. Those clouds then move inland, far from the ocean. This is why rainfall occurs far from a coast. Without this tree-caused water cycle, most of the Earth would be a desert. That is why Australia now has only a small band of fertile land along its coast. Early settlers severely deforested the land, creating inland deserts over much of the land base.[14]

A tree can put up to 250–400 gallons of water into the air in an area the size of an acre. Without evaporation from trees, there would be little inland rain.[15] In the Southwest, where I live, before it was heavily deforested, as recently as a few centuries ago, there was little desert and the land was much greener and more productive. There were vast grasslands with trees, far more rain, and very little runoff. The trees were cut to make ties for railroads and to support the movement west. The land was over-grazed, and now most of it is desert.

Without a coastal forest, the water cycle most of us rely on would not work. Rainforests around the world transfer water to the landlocked interiors. In parts of the Amazon, this water cycle can reach a hundred miles inland and affect rainfall for a thousand miles. Cut the trees and you create desert. That is a critical reason why clearing rainforests is suicidal. As of late 2020, deforestation of the Amazon surged to a twelve-year high. Many parts of the Amazon are drying out.[16]

So, plant blindness is a dangerous syndrome. Trees and plants are the super species of this planet; without them, very little life would exist. How can we be blind to something so basic? Our very survival depends on seeing with a greener lens.

A recent article by Fred Pearce says this water cycle creates giant rivers of water in the air that can bring rainfall as far as a thousand miles away. Without it, the Nile would dry up, there would be no Asian monsoon, and the fields from Argentina to the US would be dry and lifeless. There would be no bread in the breadbasket the world depends on.[17]

This moisture released by trees is also very cooling, and it maintains an equilibrium in many ecosystems. A single tree, Pearce says, has the daily cooling effect of two household air conditioners. Trees do this by releasing moisture from their leaves. Pearce says studies estimate that the Earth's trees and plants recycle forty-eight cubic miles of water every day. He says that a tenth of that is released by the Amazon rainforest alone.

Early on, plants helped to create the soil that makes up much of our planet's landmass. Before there was soil, all plants and animals lived in the oceans. Then oxygen levels slowly increased. This was due in large part to algae. Algae are not plants; they are a type of bacteria. But they photosynthesize and release oxygen and absorb CO_2. With more oxygen available and less CO_2 in the atmosphere, life could venture out of the sea and colonize the land. While early plants helped to create soil, there were lots of microorganisms helping, too. To get the nutrients they need, plants would secrete acids to dissolve rocks, releasing the minerals they needed. Later, when they evolved roots, they began physically breaking up the rocks. According to journalist Wynne Parry, plants essentially engineered much of the Earth: they caused changes in the rivers, they held banks in place, and they added enormous amounts of organic material to the emerging soils, creating diverse ecosystems. They continue to hold soil in place in every bioregion so diverse life forms can prosper.[18]

When trees are lost, there is widespread erosion. Costa Rica loses 860 million tons of topsoil each year. Madagascar loses so much soil that its rivers run red. When forests are cut, the increased sediment is dumped into rivers and streams, where it smothers fish eggs, harms reefs, and dramatically affects fisheries. One study showed that Java, an Indonesian island that is heavily deforested, was losing 770 million tons of topsoil each year, at the cost of 1.5 million tons of rice—enough to feed eleven to fifteen million people.

There are many other benefits provided by trees and green spaces. Trees help us conserve energy by shading our homes and buildings. Apartments with green space nearby have fewer crimes. A belt of trees reduces noise pollution from highways by up to ten decibels. Trees provide habitat for many other species of plants and animals. And green spaces promote a sense of community and civic pride.[19]

We cannot exist without trees and plants—but they would be just fine without us. We are responsible for the clearing of over thirty percent of the planet's forests. In recent decades, we have destroyed seventeen percent of the Amazon rainforest.[20] The Amazon was once a reliable carbon sink—one of our largest. Now, many parts of the Amazon are emitting more carbon than they are retaining.

There are about 391,000 plant species that have been identified, most of which are flowering plants. The report "The State of the World's Plants," released in 2016 by researchers at the Kew Royal Botanic Gardens, says that twenty-one percent of all plants are threatened with extinction. Many of the plant species we share the planet with are used by us and other animals in one way or another. Almost 20,000 are used in medicines. Over 12,000 are used as materials. Almost 10,000 have significant environmental uses. About 5,000 are used as human food and 4,000 are used as animal food. Very few plants are legally protected, and researchers have identified over 1,771 plants that desperately need conservation protection.[21]

It is baffling that we could be so unaware of our essential multi-level reliance on plants. It is horrifying that more and more people have plant blindness or simply do not care. This degree of anthropocentrism may well prove deadly.

It seems that things really started to go askew about 10,000 years ago, when humans developed agriculture and settled into stationary communities. Before that time, humans were foragers. Their lives were nomadic. They lived in small, closely-knit cooperative groups. The outer world had dangers, and our ancestors were filled with considerable fear, but they still had the benefit of enchantment and wonder. They were still fully immersed in nature. It was how they survived. Their minds were attuned to the total environment—the whole, not just the parts. Then, somewhere in the recent past, we learned to compartmentalize. We also learned to see the world in linear way. The world is not naturally compartmentalized or linear. This is not the lens our hunter-gatherer ancestors saw things through, nor is it the way indigenous people see things. Compartmentalization and linear thinking were aberrant abilities that required training.

As journalist George Monbiot explains in his book *How Did We Get Into This Mess?*,[22] compartmentalization and linear thinking are not innate behaviors. Young children, he reminds us, do not see and act in a linear manner. The minds of children have to be conditioned to conform to many cultural expectations and perceptions. Children are not naturally linear. They wander off in the direction of whatever attracts their attention, and we can barely keep up with them. As Monbiot says, unlike us, they do not begin a task with a view of its conclusion, but they go for what attracts them and are engaged with it for as long as it is exciting to them. They then move on to something more interesting. Most animals also behave that way.

Hunter-gatherers covered a wide territory. They did not use straight lines and a grid formation to find their dinner, although they may have had elaborate and detailed spatial interior mental maps to guide them. They stuck with a task as long as it was rewarding and then moved to another location to continue their search for food. They perceived their environment spatially and holistically—that is, from the perspective of what is happening or existing within a space as a whole, not in its parts, lines, and boxes.

But with the development of agriculture and farming, humans learned to think differently. Monbiot speculates that these innovations in our perspective developed because to grow food rather than forage, our ancestors had to follow a plan from one point to another across weeks and months. They had to plow in straight lines, make linear trenches. Soon, he says, every aspect of their lives, including their communities, was lived in grids. Linearity and management took over their lives, and they boxed themselves out of the natural world. This was just part of the new cognitive skill set they developed.

It also means their brains changed. They had to develop new neural pathways to effectively use new functional skills. Even though children are thought to go through stages of cognitive development that are believed to gradually unfold, they need to be trained to act in linear and compartmentalized ways: color between the lines, walk in straight lines, stay on the sidewalk. These skills are important aspects of our social development and our Western worldview, but they are not genetically coded behaviors. In other words, there is not a blueprint that unfolds. There are many influencing factors. We are adaptable organisms. We had to learn to separate ourselves from nature. We had to learn how to limit our perception and to see in parts, not wholes. None of this was normal or natural. And these perceptual changes actually altered how we experienced reality. But the reality of nature did not change—we changed.

While we lived in communities that required social skills and collective behaviors, that was only part of the story. The ability to compartmentalize our reality became important, along with other new cognitive skills, such as dualistic thinking and reductionism. This skillset led to major societal advances and unprecedented growth. We learned to take pride in ourselves as rational beings, set apart from all other species. Anthropocentrism was not just God-given; it was, we told ourselves, a sensible, reasonable conclusion. Of course, we did not fathom that it was the only conclusion our relatively dissociated minds could harbor. These new learned behaviors and perceptions led to the development of modern science, architecture, medicine, engineering, finance, and modern warfare. These skills helped civilization as we know it evolve. They allowed our species to leap forward at an unquestioned, unchecked, frantic speed.

It is now time to both question and restrain some of these learned behaviors and perspectives. Consider the pursuit of individualism that arose from the Age of Enlightenment and birthed the concept of the rational man. It continues to be the mantra of contemporary economic growth, leading to the pursuit of endless progress, whatever the cost. Unconstrained capitalism and market forces have now trashed the very fabric of life that we were supposedly the custodians of. And for what purpose? Surely we are not happier, wiser, or kinder. But we certainly are more neurotic.

For ultimately our big disconnect required the development of subconscious coping behaviors or defense mechanisms such as rationalization, disassociation, isolation, and alienation, as well as other common neurotic behavioral patterns.[23] These coping mechanisms became essential because we lost our sense of connection to something larger than ourselves; we were no longer integrated within the fabric of the whole of existence. We could no longer perceive holistically.

Our belief systems became fragmented and dualistic, and so did we. We may not comprehend that these responses arose out of an inner conflict or misunderstanding, but sadly they did, and as renowned German psychoanalyst Karen Horney suggested decades ago, they are part of the neurotic personality of our time.[24] We lost our footing. The alienated modern mind made the outer world, encompassing all of nature, impersonal and machine-like. This was, of course, a projection of our own gradual inner disassociation. In other words, we modern humans got very lost in a maze of our own creation.

The stories we humans told about ourselves, our old mythic framework, no longer made sense. We lost our way. It all happened very, very fast, in a matter of a few centuries—a mere drop in the river of evolutionary time. Now our scientific elite is telling us we are on the verge of dramatically altering, in fact destroying, our planetary home, possibly causing our own extinction. We have lost sight of the fact that we even have a planetary home, much less that we are naively and thoughtlessly destroying it. We have been living in an alternative reality, disconnected, disassociated, and disillusioned. The alarm is sounding, and the wake-up call is sudden and shocking. And many are still in denial.

This is where our singular brilliance got us. The challenge now is, how do we go back to a safe and sane ground? Or rather, how do we escape the abyss we are blindly running toward? We are not only rapidly destroying our life-support system, we are also on the verge of drowning in the psychic debris of our own creation.

Dualism, compartmentalization, linear thinking, ego aggrandizement, and objectification of nature are all learned behaviors that form mental constructs or patterns in our brains. Like plant blindness, they are cognitive biases. These constructs are filters that alter and limit our perception of reality. Like all learned behaviors, they have become habitual. These ways of perceiving have predominated because we have created new neural pathways to accelerate their functioning. But since they are not innate or genetically coded, we can learn to revise and modulate how we use these behaviors. Becoming aware of our cognitive biases is a first step.

Cognitive neurologist Michael Gazzaniga explained many years ago that the brain has amazing plasticity. It is not fixed; it regularly develops new neural pathways. It learns.[25] The brain has an amazing ability to adapt. It changes as a result of interactions with our environment. Every time we learn something new, we create new connections between our neurons. Want to take up golf? Your brain will soon develop new neural pathways. The more you practice, the better you will become, all because your brain will continue to develop more neural connections and pathways and change how its circuits are wired. Each lesson connects new neurons and changes the brain's functioning. That is why we can get better at certain skills, if we work at it.[26] So take up juggling, learn a language, study meditation. Your brain will change. Since we have the ability to change, why not learn new behaviors and perspectives that are less self-absorbed, less linear and compartmentalized, and more holistic, environmentally conscious, and sustainable? Why not learn to see green? Your brain will change. These greener adaptations may be essential for our survival. As we have seen, plants demonstrate green brilliance; maybe we can develop our own form of green brilliance as well. Our survival may depend on this greener, more conscious lens.

It is important to remember that our self-importance has also had a few other wake-up calls in the past centuries that have shaken us from our self-absorption. Freud is credited with pointing out that two big blows to humanity's self-perception were when Copernicus told us that we were not the center of the universe and when Darwin told us we were merely animals. This was nothing we wanted to hear, and we deeply resisted it—yet ultimately, we had to alter our self-perception. Freud also told us that what we think of as objective reality is false; it is simply determined by the subjective reality of our unconscious.[27] A frightening thought at best.

The point is that we have the psychic resilience to endure. We can absorb new information that dramatically shifts our perception. So, the revelation that we are actually a participatory part of a larger unfolding whole should not be an impossible notion for us to assimilate and run with. We desperately need a new vision, and this one has always been there. We have just not been able to grasp it because we have been so lost—so confused. We must stop listening to the collective mantra of denial, telling us we are destroying our life-support system, but we can't seem to stop ourselves. We must embrace a new vision.

In a way, our confusion is understandable. Our evolution may have happened too fast. The cognitive leap we took some 30,000–50,000 years ago, whatever its cause, was not a gradual, natural process. Evolution does not normally work that fast. Change until then happened gradually. But with this leap, there was no time to adapt, or to adjust. This was an unprecedented leap that did not leave our species time to integrate or modify the changes it brought. Since then, the rate of change has only accelerated. And there has been no external monitor. No self-correct function. No escape button. No undo.

This is it. There is no planet B as the wiser and younger among us are shouting. Maybe we will hear them.

The way out is the way in. If we soften our cognitive biases a bit (and it takes practice), we might begin to perceive that there is no separate human subject. The world is interconnected. The nature of reality is not separate—it never has been. The consciousness of nature pervades everything, and human consciousness in all its singular brilliance is merely an expression of nature's essential expression. From this perspective, we can begin to comprehend the deeper reality of the world we live in and are a part of.

Even though we are the main cause of the current planetary crisis, it is clear that the human form and purpose is not a distortion or aberration. And it should also be clear that we hold the solution to the crisis we have created. We are an essential part of the whole. We may have been confused by dualism and our self-importance, but those errors are correctable. When we wake up and shake off the distorted dream we have been captured by, we can harness our powers and direct them to the task before us. We can self-correct. We are a not a fixed species; we are a species that is rapidly evolving.

So, we need to take a deep breath, to draw deep, just as the roots of the tree draw deep into the rich and potent soil. We need to draw from the rich, collective potential of the soil that nurtures the seed of all conjecture and myth. We need to grow our new vibrant stories—our new, green-actualizing myths. The old ones are dead, mechanical, flawed, and can no longer lead us forward. We need new cultural perceptions, and they are evolving. It is through these new cultural perceptions that we will evolve.

This process has already started. Sustainability. Eco-centrism. Gaia. Integral ecology. A new paradigm is emerging that tells us that nothing can be separated from anything else and that everything is conscious. These new stories not only point the way forward, they also bridge the past. We are not alone. We are not machines. We are not abandoned. We are emergent. This is a new birth. We have only been temporarily confused, but that is understandable. Forgive and move on. Embrace what stands before us. The new paradigm is born from the collective psyche—from the heart of scientific reality. An archetype is a universal human pattern. The new archetype—the way forward—tells us we live in a holistic and participatory world. The alienated Western mind is breaking free of its self-imposed constrictions. Make no mistake, this is a true birthing process, with all its pain, gore, and glory. From this psychic rebirth, we will emerge into an integrated, participatory unified consciousness, reconnected to the whole we never really left.

Thomas Berry, Catholic priest and evolutionary historian, tells us that the evolving universe, of which we are a part, is the new guiding green story for our time. It is the story that has arisen from modern science and empirical observation. This is not gooey, New-Age thinking. The hard, scientific facts tell us that we cannot understand the universe we live in until we understand how it functions as a whole. Reductionism must be complemented by integration. In this new story, science and the material world and the more ephemeral spiritual dimensions come together. The concept that the entire universe, in communion with all parts and the whole, is, he states, widely recognized by contemporary scientists.[28] That means our mind and our intelligence are a part of the universe.

This narrative tells us the universe is not merely mechanical. It is consciousness itself. This means we humans and the other beings that we share the planet with are all conscious. Consciousness, therefore, must be a dimension of reality.[29] This new origin story is the archetypal narrative for our time. It may surprise you to learn that it is actually what science is pointing to, albeit with reluctance.

The mathematical cosmologist Brian Swimme and Berry tell us the universe from the beginning has always had a mysterious, self-organizing quality. There is nothing mechanical about it. Swimme and Berry suggest, as did Saint Francis, that everything exists in intimate interrelationship, especially all living beings on the Earth.[30]

We cannot progress in understanding this new narrative until we put to rest a few assumptions that have tainted how we view consciousness. We need to look at what scientists are now telling us and come to our own conclusions. Among scientists, there are currently several theories of consciousness. Neuroscientists have looked for evidence of consciousness in the brain but have not found it. Consciousness does not seem to lend itself to reductionist approaches. It does not appear as though it is a biological state generated by the brain. As Nicholas Christakis pointed out, "it is not a cake mix." We can't just throw together the ingredients for life or consciousness, stir it up, and bake. It doesn't work that way. One theory, called the integrated information theory (IIT), developed by neurologist Guilio Tononi, starts with consciousness and attempts to work backward to understand the physical processes that give rise to consciousness. This theory is based on the idea that conscious experience requires the integration of a wide variety of information and is simply irreducible. Somehow, he says, the brain weaves this complex web of information together and consciousness emerges. While this approach is more holistic and does propose conditions for consciousness, it still does not answer the question. It tells us that consciousness emerges, but it does not say why or what it is.

Another theory is that consciousness is like computer memory—but is it? In fact, it may be nothing like machine-based processing.[31] IIT implies that all systems that are adequately integrated—that is, systems that are complex—have some minimal consciousness. That would imply that all animals, all bacteria, and even individual cells have consciousness.[32] Yes, that could include all plants as well.

The fact is that we do not know why or how consciousness arises. We cannot find a physical mechanism. We can't get there through inductive reasoning. And it appears that we even have to look beyond physical material reality to understand its origin. There is a new story evolving. That is what many scientists are implying.

If there is not a physical quality behind consciousness, then what is it? Could it be that consciousness is a nonphysical property that emerges from the same things that give rise to physical properties? This theory, proposed by David Chambers, New York University professor and cognitive scientist, says that conscious properties are fundamental aspects of the universe, like electromagnetic charges, and they interact with physical qualities. So it is possible that it is more like a universal force than anything material.[33]

As we have noted, most scientists have only considered consciousness as a material quality, and that concept does not appear to work. That is why some scientists are now cautiously considering whether there may be more than a material aspect of the universe, as Chambers suggests. John Archibald Wheeler, who is considered one of the most eminent twentieth-century physicists, boldly proposed that reality might not be wholly a physical phenomenon. Theoretical physicist Bernard Haisch stated that if a system has enough complexity to create energy, it could generate consciousness. Physicist Gregory Matloff of the New York City College of Technology suggests that, although speculative, the concept of panpsychism (see below) might explain dark matter. Christof Koch and Guilio Tonori, both leading neurologists, carefully suggest that, based on their work, consciousness is an intrinsic, fundamental property, is graded, and is common in all biological organisms, even simple biological systems.[34]

If all matter is imbued with consciousness—and this is still speculative—that means all animals and all plants have some type of consciousness, since consciousness is theoretically a fundamental aspect of the universe. The theory of panpsychism says that all aspects of reality have some psychological properties in addition to their physical properties. In other words, everything innately has an element of individual consciousness, and this quality is nonphysical. This theory says that the entire universe and all of nature is conscious.

As explained by David Skrbina in his book *Panpsychism in the West,*[35] panpsychism is based on rationalism, empirical evidence, and evolutionary principles. Everything in the universe has the same building blocks, so consciousness is intrinsic in everything. This may be the simplest possible explanation for what consciousness is and how it arises.

This also means we have to get beyond the idea that consciousness requires a brain and nervous system like ours. Adam Frank, who is an astrophysicist at the University of Rochester, has spoken about the reluctance of scientists to suggest that we can even consider that consciousness can exist without the brain.[36] Of course, this does not imply that human consciousness and the consciousness of a worm, a single-celled organism, a robin, or a plant are the same. It merely says that everything has some innate consciousness. He suggests that matter and consciousness are entwined and that consciousness might be a new entity entirely and not contained in the law of particles. He also suggests that consciousness might be an aspect of what the entire world is built of.[36] Skrbina says that it is our human bias, our anthropocentric perspective, that we need to overcome.[37]

We are being asked to consider a new narrative, a new perspective on creation and on reality. This is what Thomas Berry suggested. The universe and all the parts that make up its whole are conscious. Consciousness is innate in all of matter.

Deepak Chopra, a Hindu practitioner and medical doctor and author of eighty-five books, has long advocated for the integration of science and spirituality. He states, "Consciousness is the universe. And consciousness is universal. It is everything, always, and everywhere."[38] He says we can and must liberate ourselves from the conditioning and mental constructs that underlie our anxious and ego-driven lives, our false reality. He tells us that the new creation story is, simply, consciousness. His most recent book, *Meta Human,* sets forth a new spiritual perspective, based on science, that redefines our role in the universe. You may not agree with his conclusions, but his book is endorsed by dozens of scientists across many disciplines.

So, what if science ultimately does conclude that everything is conscious? What if we actually are part of a larger interactive whole? How does that change things? It is not really that big of a stretch. It is just a shift in perception. It means we are individuals, but individuals within a larger whole. It means we have both the ability to perceive in a linear, reductionist manner when that is useful, and the ability to perceive more holistically. This wider perceptive gives us the ability to function more fully. We are in fact not just singularly brilliant; we can become fully functional as multidimensional brilliant beings. The new paradigm and the new creation story that comes with it define our new birthright. We can perceive holistically because we are holistic; we are part of a whole system, of a whole interactive network. We have the ability to perceive as a system, not just a part of a whole network. Trees and the entire plant world seem to do just that already. They already have a green consciousness.

We might be closer than we think to perceiving with multiple, greener awareness. Consider the widely used term "sustainable." Most people understand that for life to continue, we can't deplete resources faster than those resources are naturally replenished. A balanced environment requires an understanding that everything is connected. Sustainability requires a holistic, systems-based approach, where the economy, the environment, and cultural, technological, and political aspects of life are integrated. It involves concepts like ethical consumerism, green building, renewable energy, and essentially applies to all aspects of human endeavor. The term has been mainstreamed and is widely used. Communities across the world have adopted sustainability goals. In 2015, all the United Nations member states adopted the 2030 Agenda for Sustainable Development. The Green New Deal is a set of sustainable policy proposals in the US that would address the climate crisis.

The Gaia hypothesis suggests that the Earth is a living system in which all organic life and inorganic life interact in a synergetic and self-regulating manner. Our planet is an integrated whole. The nature of life is holistic. We used to think that life adapted to planetary conditions. We now understand that it is the whole system that does the regulating. This does not contradict Darwin's concept of natural selection, it expands it. As stated by Lovelock in 1986, Darwin's vision is extended to include the largest living organism in the solar system: the Earth itself.[39] The system works to keep conditions on the planet within boundaries suitable for life. Or at least it did until recently.

Consider the term "integral ecology," introduced by Pope Francis. Integral ecology is considered a new paradigm of justice, both in how we treat the environment and how we treat other humans. The Pope says that nature is not separate from us. Everything is connected. The term re-visions the relationship between human beings and the natural world. Integral ecology tells us that humanity is integral to nature and dismisses the notion that nature is subject to human domination.

None of these terms are difficult to understand. Most of us get that we need to live sustainably, that the Earth is a living system with natural feedback loops, and that, like all living things, the Earth will try to maintain balance and equilibrium. We can understand what the Pope is telling us, that we are not separate from nature and that everything is connected. We already grasp this. The harder part to accept is that there is no major role for anthropocentrism in the new story. It is not a story in which we have the lead role. We are only supporting actors in a cast of millions.

While anthropocentrism may be on its way out, I personally am not yet comfortable with the term ecocentrism. This term is used to define an ethical approach that is supposedly nature-based, rather than human-based. It implies that we humans have no greater value than any other aspect of nature and that there are no existential differences between humans and nonhumans. The term was allegedly first used by Aldo Leopold, but I am not sure that this is what Leopold meant. His book *Sand County Almanac* was my bible for many years. Leopold believed we have a responsibility to care for and respect all of the natural world and developed the concept of the "land ethic." He believed we are a part of the community of life—a community of interdependent parts that includes all of the Earth.[40] That would include the soil, water, and all three kingdoms of life: plants, animals, and fungi.

The problem I have with ecocentrism, as it is currently defined, is that even though it purports to be a system of values, it offers no clear guidance. I couldn't buy anything in a grocery store if I practiced ecocentrism. How would I decide what was an ethical decision or ethical consumer choice if everything human and nonhuman is equal? There are almost 8 billion humans. What would we do with major population centers like New Delhi? How would we provide food, water, and shelter equitably for ourselves and all other species? Ecocentrism may be an aspirational philosophical concept, but it is does not offer the pragmatic value or moral guidance we need to make clear choices.

I can grasp that we are all part of a community of life, that we are part of a whole, and that we live in relationship to the whole, rather than separate from it. Maybe we don't need anything in the center. I like the approach suggested by animist scholar Graham Harvey.[41] He suggests a new definition of animism that is neither anthropocentric nor ecocentric. He suggests we learn to see the world as a community of persons, most of whom are nonhuman species, but all of whom are related and all of whom deserve respect. The approach, he suggests, which builds on many traditional indigenous belief systems, tells us we are relations with all other beings and that the rights and responsibilities of all members of our multispecies community need to be taken into consideration. If you do not like the term "relations," substitute the word "connected." This approach does not deny that some of us are predators and some are prey. That is the nature of all life and all interaction. We all need to eat.

The relationships and responsibilities we have with all other members of our multispecies community are not easy, nor are they nonviolent or bloodless. Harvey paraphrases a shaman who advised that "in order to eat it is necessary to consume relations (as there are no other kinds of beings)."[42] That, very simply, is it—and it is an old dilemma. But relationship and responsibility need to be viewed in tandem. Harvey says we have to understand that consumption has a relationship component. It also has a responsibility component. We have to shift our thinking to include the idea of interspecies respect and gratitude in our consumptive patterns. Our consumptive patterns and choices should include respect and gratitude for our relations, for all the other beings that are a part of our food chain. That is our core responsibility.

Our consumption patterns are learned behaviors, like all social behaviors. Perceiving in a more holistic manner is also a learned behavior. Sustainable agricultural practices and providing humane conditions for domestic animals are sensible and easy to implement. As consumers, we have choices and can and will demand change. What about plants and trees? At minimum, we have to unite to stop all further rainforest destruction. This is not optional. That means boycotts of all products like Brazilian beef and soybeans until that country stops their destructive practices. Palm oil plantations are rapidly destroying rainforests in Indonesia and Malaysia. Palm oil is not essential for our survival, but rainforests are. Numerous agricultural processes are sustainable, and climate change is incentivizing a new global wave of essential practices and policies that are sane, practical, and will dramatically change how and what we consume. These practices will hopefully lessen our wasteful, thoughtless behaviors.

We need to understand the interconnection between our consumptive patterns and environmental destruction. We cannot afford plant blindness, nor can we afford mindless consumerism. But it goes beyond that. If we understand, even in a small way, the web of life we are a part of, we will make thoughtful choices. We will pay attention. We will respect the rights of an old-growth forest or of a riparian ecological system. If we fully understand that we could not exist without trees and the diverse plant communities that cover the planet, we will comprehend that trees really are the super species. We not only have to be grateful for their existence, we must do everything possible to keep their fragile ecosystems intact. Their survival is our survival. That is what seeing green means.

Even if we are slow to comprehend that everything is connected, interactive, and conscious, as science and religion are now suggesting, we will gradually awaken. The circle of our awareness will get bigger and more expansive. We will innately understand that we are a part within a larger interactive whole. This is the new world view—the new paradigm. It is not that big a leap. It is a view that also tells us that every aspect of our lives is sustained by the plant kingdom in some manner. We will gradually learn to see through a greener and more humble lens.

7
Restoring, Rebalancing, Regreening

Spruce and Fir (Picea *and* Abies)
For the indigenous Siberians of the Altai mountains, the World Tree is a gigantic spruce that reaches from the navel of the Earth to the highest region of the heavens, thus connecting the three main layers of the universe: the spirit world, the earthly plane and the underworld. . . . The spruce is also the centre of sacred teachings among the indigenous tribes of southern Canada, who call it the Peace Tree. Its lessons are: to cooperate and exchange with other life forms; to be connected with all Earth and the heavens; and to exhibit a joyful and tranquil humility.

THE MEANING OF TREES

We have learned so much in recent years about trees, forests, and natural environments of all types. There are legitimate scientific debates about whether trees and other plants communicate, are intelligent, have consciousness, and are sentient beings. Those are fascinating questions, and while they may not be resolved for some time, they are causing us to pause. We are being asked to reevaluate our perceptions and understandings of how we relate to and value the other species we share the planet with. This shift has many aspects. We now understand that many animals have consciousness. No one would seriously question that our nearest ancestors, chimpanzees, who share ninety-nine percent of our DNA, have consciousness and are sentient beings. We now know that many animals are intelligent, can learn, can remember, and adapt. Dolphins show creativity. Elephants are social, help each other, and have amazing long-term memories. They mourn their dead; they grieve. Crows and all corvids solve puzzles, do math, and recognize themselves in mirrors. Many animals have self-awareness and show signs of altruism. It has been a slow process, but we are gradually coming to understand that we are not the only ones here with feelings, emotions, intelligence, and some form of consciousness. Because of this new understanding, we have moved in the direction of granting some rights to other sentient beings, at least philosophically, and that may at some point include the plant kingdom.

Prompted by the climate crisis, new clarifications are emerging from our religious systems. We do not have dominion over other species. We have a duty toward all living beings. We are the stewards, the custodians, and the caretakers of planetary life. We are not just usurpers—not just consumers. We have a role, a value, but there is nothing supreme or special about us other than the supreme mess we have made of things due to our misunderstanding and arrogance. We have a duty to set things right. And yes, both our spiritual progress and our survival depends on it. But that is not the only reason we need to radically adjust our behaviors, mindsets, and priorities, and work wholeheartedly to restore, replant, reclaim, and rewild our fragile planet.

We have been given everything by our green allies: the air we breathe, the food we eat, and the medicine that heals our bodies and balances our minds. The hard truth is that they could easily live without us, but we could not last without them. They offer new insights about community, living, interactive systems, and network responsibilities. They model different forms of consciousness, intelligence, and holistic behavior. We really do need to be more like trees. We have a lot to learn from them. As we are beginning to understand, nature as a whole is a powerful, spectacular teacher.

But first, we have to stop the wave of planetary destruction we have set in motion. It is not too late. People around the world are starting to wake up. In the spring of 2018, the Pew Research Center evaluated how people globally see climate change. Eighty percent or more of the people surveyed in Greece, South Korea, France, Spain, and Mexico saw climate change as a major threat, and only four or five percent of those surveyed thought it was not a threat. About three-quarters of the folks surveyed in Japan, Argentina, Brazil, Germany, Kenya, Italy, and the Netherlands saw climate change as a major threat. Sixty-six percent of Canadians and Australians said it is a major threat. The US came in at fifty-nine percent. Russia and Israel were the laggards at about forty percent. In almost all countries, those who saw climate change as no threat at all were in the single digits.[1] Since 2018, those numbers have only increased. Seventy-six percent of those who live in India now believe it's a major threat. We can no longer ignore it. It is in the news every day: massive fires, ocean dead zones, coral loss, unrelenting flooding, and soaring temperatures. Climate catastrophes are mounting all around the world. And almost everyone now is connected by the internet and is watching it happen.

We must also remember that most people worldwide are good and caring individuals. We care for our families, offspring, friends, communities, social networks, and animal companions. We love our forests, our parks, and our gardens. We are social and cooperative by nature. We strive to be kind and altruistic. We love our planet and we will work to help it through the crisis we are in the midst of. We also are capable of learning and changing. That means not just millions but billions of people will be activated and will demand change. And fast change is required.

Our planet is on fire. The link between global warming and wildfires is now clear. As temperatures warm, they dry out fuels, resulting in longer, hotter fire seasons and more devastation. Australia is a climate catastrophe. Forty-six million acres burned just in the early part of the 2020 fire season alone. A billion animals are believed to have died in those fires. Eighty-five percent of Australia's plants and animals are found nowhere else, and many will become extinct. The fires will change water flows and vegetation and will dramatically increase carbon emissions and further devastate the Great Barrier Reef. The tipping point came more quickly than anyone predicted. Nerilie Abram, a climate researcher at Australian National University, explains that drought and loss of forests cause higher temperatures over the land and lower humidity. A shift in the winds, believed to be caused by global warming in the southern hemisphere, reduced rainfall, causing a long-term drying trend, lower humidity, and unprecedented, fierce fires. With each degree of warming, this trend will severely worsen.[2]

Australia has also been recognized as a global deforestation hotspot—the worst of any developed country. In recent years, they have bulldozed a massive amount of native woodlands to create more land for livestock. Projections are that by 2030 they will have destroyed three million hectares—about 740 million acres. They are destroying woodlands at the rate of 1,500 football fields a day. To give you an idea of how massive that is, consider that all the farmland in the United States is a little over 915 million acres. With that much tree loss, of course they have changed their climate, and possibly the world's climate, and have fueled massive fires. The connection between tree loss, warmer temperatures, and climate change is well established.[3]

I think we really need to take in the magnitude of these losses. We need to feel it. I am reminded of a personal loss that occurred when I was younger. It has stayed with me. There was a lovely hardwood forest across from my aunt's woodlot. It was dark, lush, rich, and moist, with a full canopy. I spent a lot of time there when I was young. It was filled with magic. There were red, yellow, orange, and bright white fungi everywhere. Every time I visited, I found something new and exciting to explore. There was phosphorescent foxfire that glowed in the dark. There were berries and plants I had never seen before. Frogs and butterflies. Then, one day, I came home from college and it was gone. It had all been bulldozed by our neighbor to make new pasture. There was nothing left but rotting piles of uprooted trees and soil. I was devastated. This was just a few acres, but it helps me to remember this personal pain and loss I felt when I hear and read about the vast devastation that is occurring across the planet. I need to take it in, to be aware of what is being lost, so I can act to prevent more losses. Our pain can wake us up.

Recently, there have been major fires in the Artic, Siberia, California, and the Amazon. Due to increasing temperatures and long-term droughts, California has become a tinder box—and will be for years. In 2018, over eighteen million trees died, increasing the total to about 150 million dead trees awaiting fire. The old norm was that about a million trees a year would die from insects and diseases. Now, heat, drought, and bark beetles are causing unprecedented tree deaths.[4]

Even the iconic giant sequoias are dying. This is shocking the forest ecology specialists who study sequoias because these giants were thought to be indestructible. They have lived through all other fires and droughts and were believed to be immune to beetle infestations. These amazing trees normally live to be 3,000 years old. Now, due to extreme drought, they are being severely weakened and are then attacked by infestations of beetles that are slowly killing them.[5] And the high intensity California fires are now killing thousands of these irreplaceable giants. An estimated 10,000 died in the 2020 fires. Over 2,200 died in fires in 2021. We have lost up to nineteen percent of all sequoias in recent years.

California lost 1,893,913 acres to fires in 2018 and 253,214 acres in 2019. But 2020 broke all records, with 9,387 fires burning 4,359,517 acres—a shocking four percent of the state's land. By the fall of 2021 another 2,500,000 acres had burnt. In Colorado, massive infernos raced through the Rockies and other areas, raging at the unheard-of pace of 6,000 acres an hour. In 2022 after extensive droughts, massive fires hit New Mexico and fifty square miles of forest and villages just east of my home were destroyed. In southern New Mexico, even larger fires brought our state's annual total of lost forest to 904,422 acres. And further west a pattern of atmospheric rivers dumped unprecedented amounts of moisture, leading to massive flooding, mudslides, and destabilization of many forest communities. These new patterns of drought, fires, and flooding will only increase throughout the world.

The Amazon in Brazil is very vulnerable. In 2019, there was a 300 percent increase in fires; most were human-caused. We have already deforested seventeen percent of the Amazon. Many Amazonian trees evolved with thin bark and cannot survive fires. It is estimated that it will take thousands of years to regain the carbon storage they provided.[6]

In addition to trees lost due to drought and infestations, according to The Rainforest Action Network, we lose between 3.5 billion and seven billion trees a year through other means. Of that loss, timber harvesting makes up thirty-seven percent, agricultural expansion twenty-eight percent, wildfires twenty-one percent, and roads, mining, and infrastructure about fourteen percent. Some sources say we are losing far more trees—as many as fifteen billion trees a year. This results not just in loss of plant and animal diversity; it also increases CO_2 emissions by over twenty percent.[7]

We have about a trillion trees left on the planet. We know this because they have been counted using satellite technology. We believe there were once three trillion. Tom Crowther of the Swiss University ETH Zurich, who has led much of this research, estimates that there are about 1.7 billion hectares (a hectare is 2.4 acres) that we could grow additional trees on. Their research estimates that a worldwide planting program could remove two-thirds of all human CO_2 emissions. It will take fifty or more years for new trees to grow large enough to effectively remove carbon. There is not time to waste, and it is not the only solution. Restoration in the US costs about thirty cents a tree. We could plant a trillion trees for about $300 billion, Crowther says. That is not that much money. According to CNBC, the US has spent over $6.4 trillion on the wars in the Middle East since 9/11.[8] And in 2017, according to the International Monetary Fund, fossil fuels got $5.2 trillion in global subsidies and the rate of subsidy has not yet changed significantly.

Forests cover about thirty percent of the world's land. However, we have a net loss of about ten billion trees a year now. Crowther's research is based on measurement of tree cover conducted by hundreds of people using 80,000 high-resolution satellite images from Google Earth combined with AI computing of key soil topography and climate factors to map where trees could grow. Of the land Crowther identified that could support new trees, about two-thirds has trees already, 3.2 billion hectares is treeless, 1.5 billion hectares is used for growing food, and the remaining 1.7 billion hectares is potential forest land. Some is now grazed by livestock. The world's six biggest countries—Russia, Canada, China, the US, Brazil, and Australia—have the most potential as restoration sites.[9]

Many countries are on board to some degree. The Bonn Challenge, which was backed by forty-eight nations, plans to restore 350 million hectares of deforested land by 2030, but they need to restore far more and move faster. India is moving toward its goal, and in August of 2019, over a million people planted 220 million trees in one day. A wall of trees called the "Great Green Wall," about 8,000 kilometers long (4,070 miles), is being built over twenty countries. This initiative includes Algeria, Burkina Faso, Chad, Benin, Cape Verde, Djibouti, Egypt, Ethiopia, Libya, Mali, Mauritania, Niger, Nigeria, Senegal, Somalia, Sudan, the Gambia, and Tunisia. It is about fifteen percent completed, and when it is done, it will be the largest living structure on the planet. Pakistan is ahead of their scheduled target of planting a billion trees and has a new target of ten billion. The UK is planting five to ten million trees. In the US, in contrast, we plant 1.6 billion trees a year, but half are planted in plantations by forest product companies. We lose at least 1,729,376 acres of forest a year in the US.[10] That was before the 2020 fires. In early 2020, the World Economic Forum launched the Davos Initiative to unite and promote reforestation efforts worldwide with the goal of growing, restoring, and conserving a trillion trees intended to restore biodiversity and fight climate change.

Presently, the tropical regions of the planet release the highest level of greenhouse gases due to environmental degradation and destruction. Ironically, these areas have the greatest carbon storage potential. According to a report by Conversation International, Woods Hole Research Center, and the Nature Conservancy, Brazil, Vietnam, Malaysia, India, Zimbabwe, Niger, and New Guinea could reverse their CO_2 emissions, which add up to about a quarter of the world's total, by practicing basic stewardship, protecting forests, restoring forests, managing forests, and protecting wetlands. There is no question that many poor countries would need climate mitigation financing to do so.[11]

In some countries, drones are planting trees. That may sound strange, but it works. They fly across a designated area, collect data about soil conditions, and determine the best locations for planting. Then they fire biodegradable pods into the soil filled with a seed that has germinated in a special sauce of nutrients. In Myanmar, it has been demonstrated that two drone operators can plant 400,000 trees a day.[12]

This idea was dreamed up by a group of Canadian science and engineering graduates who set the goal of planting a billion trees by 2018 using drones. They call their project Flash Forest and say drones plant trees ten times faster than a single worker at eighty percent of the cost. Biocarbon Engineering, the company that makes the drones, is training local people who used to make a living making charcoal to be drone pilots and to reforest mangroves. Mangrove forests sequester an enormous amount of carbon and half of these forests have been lost. The trees they plant also earn carbon credits. Restored mangroves support aquaculture and are essential estuaries. Everyone wins.[13]

A UK tech company called Dendra plans to plant 500 billion trees using drones by 2060.[14] This is just one of many exciting innovations using new technologies, financial incentives, and local communities to restore a bit of ecological balance.

China has started to build the world's first forest city in Liuzhou. This futuristic green city prototype will be home to about 30,000 people and will have about 100,000 plant species and 40,000 trees that, together, will absorb 10,000 tons of carbon dioxide and fifty-seven tons of pollutants while suppling 900 tons of oxygen a year. It will also reduce air temperature, create habitat, and minimize noise. The entire city will be energy self-sufficient and run on renewables, including geothermal energy.[15] More efficient cities are planned not just in China but around the world.

Another innovation that is getting a lot of trees planted is the Plant-for-the-Planet cell phone app. The app has fifty reforestation projects in developing countries that you can chose from. For $3.00, you can plant a tree in Brazil. For $108.00, you can plant 1,000 trees in Indonesia, and the money goes directly to tree planters. You can even see photos of the trees you helped get planted.[16]

We know that regreening the planet can reduce emissions. Trees in the US alone remove eleven to thirteen percent of emissions per year, yet we have destroyed as much as ten percent of the planet's wilderness in just the last few decades. A study by the Proceeding for the National Academy of Sciences has found that planting trees, restoring peatlands, and improving land management could provide up to thirty-seven percent of the greenhouse gas mitigation we need to keep global warming to two degrees Celsius by 2030.[17]

The UN Convention on Biological Diversity says we need to protect up to a third of the world's land and oceans by 2030 to reverse the decline in biodiversity that puts at risk the survival of humanity. If not, we risk the loss of over one million species and the collapse of our life-support systems.[18]

Trees, which are the true super species of the planet, may also turn out to be our superheroes and saviors. Perhaps they are the archetype for our troubled age. They may save us from the worst of the climate crisis we have engineered. They may save us from ourselves. We know what they do for us and what they are capable of. If we work together using their extraordinary capabilities and green consciousness, we may be able to mitigate the worst effects of climate change.

There is a lot to learn. Not all replanting efforts are successful. Some scientists question Crowther's carbon capture estimates, but the idea of planting massive numbers of trees is wildly popular. What we have learned is that we need to do far more than plant billions of trees. Sadly, most of them do not survive. We need to also care for the seedlings, monitor their growth, and engage the local communities who live in areas where trees are planted. And, most importantly, we need to work with arborists who understand what species can be successfully introduced and what their needs are. China is a classic example.

China has planted billions of trees in recent years in an attempt to restore arid regions, some of which were depleted due to overgrazing. They started some twenty years ago in their Grain for Green program, repairing damaged farmland. Their goal was to achieve thirty percent forest coverage of their land base by 2050. They acknowledge it is a long process and there has been a learning curve. They have created water shortages in some areas by planting the wrong tree species and have had to replant with native shrubs and herbs, but huge areas are turning green. That is good for the planet as a whole, and it is stabilizing thousands of acres of otherwise erodible soil.[19]

We also now understand that savannahs and grasslands sequester a huge amount of carbon, so they are not good candidates for tree planting. We also know that the fastest way to sequester more carbon in forests is not tree planting, it's in letting degraded forests naturally recover and changing our methods of forest harvesting. Suzanne Simard has long advocated for saner forest management that recognizes the vitality of the forest community as a system and recognizes the inherent green wisdom of old growth and what she calls the "mother trees." Spare the older, mother trees, and the entire forest with its full carbon sequestering potential is rapidly restored.

Planting more trees is only part of the solution. We have lost portions of the Great Barrier Reef the size of the entire boreal forests of North America. Coral reefs are the great estuaries of the oceans. Thousands of animal and plant species have already been devastated by climate change and habitat loss. What if it is all too little, too late? We have the resources, and we have the technology to avoid the worst of climate change. But what if we cannot generate the collective political will required? The sixth extinction as we know is already beginning.

There have already been five major extinctions. According to Michael Greshko writing for *National Geographic,* ninety-nine percent of all organisms that have lived on Earth are now extinct.[20] Stop and think about that. Only one percent of all life survived all prior extinctions. One percent of all life. That is a sobering thought, but with each past extinction, over millions of years, new life has emerged and flourished. The first known extinction occurred about 439 million years ago and destroyed eighty-six percent of all life. Scientists believe this event was caused by glaciation and falling sea levels. The glaciation may have been in part caused by plants, as they were very abundant and removed massive amounts of CO_2 from the atmosphere, lowering global temperatures. The sea level changes from that period are thought to be due to the upthrust of the Appalachian Mountains.

The next extinction was the Late Devonian, about 365 million years ago. Seventy-five percent of all species died. A period of volcanism is the likely cause. This activity spewed greenhouse gases and sulfur into the air and acidified the oceans. Asteroids might have also played a role. This was followed about 251 million years ago by the Permian-Triassic extinction, when ninety-six percent of all life was lost. This one was also caused by massive volcanic activity. Then came the Triassic-Jurassic extinction 201 million years ago, when eighty percent of all land and sea species were lost. The culprits were the usual ones: climate change due to volcanic eruptions and an asteroid impact. Then came the Cretaceous-Paleogene some sixty-five million years ago, which brought an end to the reign of the dinosaurs and a loss of seventy-six percent of all species. This one was also caused by an asteroid impact. Oddly, a scrappy little shrew-like mammal that lived underground survived. We believe it was the only mammal to survive. That little mammal evolved, and that is how we got here. We evolved from the mammals that evolved from this dim-eyed rodent-like critter who rarely saw the sun. It is now extinct, but we and some 5,000 other mammals that evolved from one dim-eyed burrower are still here and facing the sixth extinction: the Holocene extinction. It has a different cause than the prior five and has an alternative name: the Anthropocene. This extinction event was caused by the activities of one species that descended from the lowly burrower. The sixth extinction may prove to be the end result of a sequence of events that began a mere 30,000 years or so ago, when an apparently random mutation occurred in the brain of our species, *Homo sapiens*. This mutation resulted in what is known as the cognitive revolution, after which our path toward the sixth extinction started, when we began to think abstractly and develop radically new perceptions. We began to view ourselves as special, separate, and the dominant species on the planet. This viewpoint justified what we perceived as our right to use and abuse the rest of nature as we saw fit. We disconnected from the whole. Over time, the natural world that had formerly been a source of wonder and awe, as well as sustenance, was commodified and minimized. We stopped seeing it as alive. We forgot that its vitality is connected to ours.

We are now on an essential course correction, and it needs to happen fast. Some say that the planet may be better off without us. But the plant kingdom is not praying for our demise. They are ready to partner with us, to help us learn to think like a community, to function as a part of a living network. We can learn to work more symbiotically with this amazing green kingdom. We can be their partners. That means we cannot just take from them; we also have to give back. I am suggesting we view our relationship with our plant allies as a true partnership. Consider it a contractual relationship—a contract between two very different but sentient systems. I encourage you to think of this idea as more than a metaphor. An emergent new contractual understanding of our relationship with other species may be vital to our continued existence.

What would a contract of this type look like? As a former contract attorney, I find that this question poses an exciting thought experiment. A contract, after all, is an exchange—a bargained-for exchange. You can't just take. That would be theft. You have to give something back in exchange. That something is agreed upon. In contract terminology, it is called consideration. Somehow, trees worked out a deal with their fungi partners. As we know, it has been scientifically proven that trees give up a third or more of the photosynthetic sugar they produce to their symbiotic fungi partners. This is an example of consideration—a bargained-for exchange. Fungi may be tough negotiators. We, however, don't just need the plant's photosynthetic sugar; we consume all their body parts in one form or the other. However, if we act thoughtfully and do not abuse our contractual rights, the plant kingdom will continue to supply enough essential resources and use its many talents for all of us and for the future generations of all three kingdoms. Acting thoughtfully means we have to think and act as intelligent actors who are part of a massive interconnected network of life. This also means we have to understand this is, in fact, the true reality.

Every contract requires a down payment. You have to bind the contract. This one is no different. The down payment for this new contract is pretty clear. We need to restore what we have harmed in our plant and forest communities, our grasslands and savannahs, our riparian systems and our oceans. We need to replant, restore, reclaim, and rewild as much as thirty percent of the Earth—that is, thirty percent of the oceans and thirty percent of our land base. And we need to stop destroying the atmosphere, stop burning fossil fuels, and stop emitting CO_2 and other greenhouse gases.

The 30 × 30 goal for diversity was set by the United Nations Biodiversity Conference. Conserve thirty percent of our planet's land and water by 2030. But not just any land. We know our forests sequester enormous amounts of carbon. Research recently published in the *Journal of Nature Sustainability* warns that we simply cannot afford to lose the carbon that is stored in our planet's tropical forests and peatlands in the Amazon and Congo basin, insular southeast Asia, the mangroves, tidal marshes, and wetlands, and in the rainforests of the northwest of North America and eastern Canada. These most critical areas and huge amount of carbon they sequester must be conserved or we stand little chance to survive.[21]

That is just the down payment. The contract itself requires that we learn to be true planetary stewards and custodians. That is our part of the deal—our contractual obligation. It always has been the deal. We just forgot we had planetary responsibilities. We misunderstood. We breached the original contract—the deal that came with the key to the garden of all life. This is our last chance to get it right. If we do, we can stop the sixth extinction—our extinction.

I don't mean to sound overly optimistic. No matter how fast we move and how much of the planet we restore and replant, we can't completely reverse the wave of devastation we have set in motion. We can only lessen its impact. Many species will still be lost. Human suffering will still increase dramatically as a result of climate change. But the difference between global warming held to two degrees Celsius or less versus four or five degrees or more is enormous in terms of our survival and that of countless other species.

COP 26 and other climate summits are based on the premise that we need to keep the climate goal of no more than 1.5 degrees Celsius alive. We do not yet have the global political will and wherewithal to save the world as we know it. A lot of really smart people worldwide are working tirelessly on the complex solutions required. The global coalition of young and terrified activists are slowly awakening the rest of us to the horrid reality we are leaving them unless we act now.

Daily, we are horrified and even sickened by the increasing planetary decline we are witnessing. It was deeply disheartening to see the priceless and effective environmental policies the US implemented in the 1970s, when I was a part of the early environmental movement, being dismantled on a daily basis by an administration of science deniers. We now can correct our course as a country, but for how long will a saner political resolve last? We are devastated as more rainforests are destroyed in Brazil after so many worked so hard to overturn destructive and suicidal policies. The level of corruption on the part of politicians and corporations seems overwhelming. However, we cannot allow ourselves to be distracted by despair. The fact is, we can still avoid a major extinction event and retain a livable planet. We have the resources, talent, and technology to not only avoid a massive disaster—we can and are creating a greener world. No one can continue to remain passive and disassociated. We have to be willing to sacrifice and to fight for the enforcement of green rights and responsibilities for the sake of our loved ones, future generations, and all species. That is what the new contract requires of us.

The green world envisioned by this contract is evolving fast. We are building a sustainable world. There are key words and concepts in this contract that we must memorize and never forget again. "Sustainable" is one. Concepts like "holistic," "interconnected," "interactive," and "integral ecology" must become our basic understandings. We have new responsibilities and new relationships to fulfill. We must respect our contractual partners. And we must be extremely grateful we are being given another chance.

If we abide by the terms of this contract, we will soon see a greener world. Billions more trees and restoration of as much as thirty percent of our land base and oceans will require heavy lifting, and it's just the beginning of the rebalance needed. Our harvests will be sustainable. The rate of species loss will lessen. We will drive electric cars and trucks; some may have their own solar panels. Already, many nations have banned gas-powered vehicles: the Netherlands by 2025, Norway by 2030, and the UK by 2035. Major car manufactures like Volvo, Volkswagen, Subaru, and even GM and Ford have said they will stop producing them. US makers are leading the charge—consider Ford's innovative all-electric F-150 pick-up truck. Solar and wind will power the world, along with other non-fossil-fuel technologies. Elon Musk, technology entrepreneur, has said that one hundred square miles of solar panels could power the entire US. That is not very much land.[22] Our buildings will be green designed and greenly refurbished, built to meet environmental efficiency standards and using sustainable materials, and will include additions like smart solar windows and functional solar roofs. Regenerative agriculture will be the norm. Our diets will be more plant-based, with a range of delicious alternatives to conventional meat and carbon-intense foods, including lab-grown meats, delicious steaks, fish and chicken-like foods made from vegetables and fungi, and air protein. Air protein is a meat alternative made from the elements found in air using hydrogenotrophs: single-celled organisms that convert carbon dioxide into protein.[23] This may become our own form of photosynthesis. Our roadways will incorporate functionality and sustainability. They will be solar and smart like the ones

China is building. Michigan will soon have its first mile of solar road. You will be able to charge your car when you drive on them. New no-/low-carbon methods of making steel and cement will be mainstreamed. Automated solar arrays that heat up to 1,800 degrees, hot enough to manufacture cement and other carbon-intensive industrial products, are being developed by Heliogen.[24] We will power airplanes with green hydrogen produced from electrolysis of water using electricity from renewable sources. We will use materials made from fungi and recycled products. Our fungi friends will help us make bionic products that replace plastics and can even generate electricity. There will be floating food farms, and solar and wind farms will dot the oceans. We will have atmospheric carbon scrubbers, chimney carbon scrubbers, smog-sucking chimneys, and algae bioreactors that capture carbon. Agriculture will be transformed and will not just be sustainable; it will lead in sequestering carbon. Any trash will be used for fuel; if it cannot, it will be recycled. This is what the future will look like, if we act soon.

I recently heard Christiana Figueres, former executive secretary of the United Nations Framework Convention on Climate Change, speak, and was struck by what she said, "We are immersed in the most amazing transition that the human race has ever started. . . . Forget gradual shifts, forget linear changes, we are in a world of exponential transformation." This is exactly where we are. We can do this.

Most of this technology already exists, and some is still on the drawing boards. We are an innovative species. But this task is not just the equivalent of a moonshot or world war. Those concepts need some reframing. There is no enemy to fight; the enemy is nothing more than our bad habits and the consequences of decisions and policies that were based on false information and false narratives fueled by corruption and ignorance. But it will be a race and already is between countries, companies, innovators, financiers, and all types of creative forces, competing to figure this out and get it right. This is where our species' singular brilliance will be put to a test, because singular brilliance will not be enough. We need to garner our collective, multidimensional brilliance and let it shine within a deeper understanding of holistic interactive systems and networks that acknowledge and respect all living systems. We are up to this task. It is a painful, challenging, and exciting time to be alive. We can do this. We can get it right. We just need to wake up and fully engage.

POSTSCRIPT

Some years ago, my nephew Bob bought my aunt's woodlot—the one I had played in as a child. He also had wonderful memories of playing in those woods when he was young and wanted his grandchildren to have the same opportunity. My oak tree is now fully mature. Perhaps when Bob takes his grandchildren to play in the forest, they will sit under its magnificent branches. I imagine them gathering acorns as I did and placing some in their pockets. I can hear Bob asking them, "What are you going to do with all those acorns?"

"Grampa," they answer, "we are going to plant them. We need to grow a lot more trees."

NOTES

EPIGRAPHS

Quotations at the start of each chapter are taken from Fred Hageneder's wonderful and highly recommended book, *The Meaning of Trees: Botany, History, Healing, Lore.*" This book is truly a delightful guide to "some of the world's most magnificent beings" and the "heroes of the forest." It was published by Chronicle Press, San Francisco, in 2005. Fred Hageneder is the chairman of The Friends of the Trees, an organization that protects trees in areas considered to be sacred.

These quotes can be found on the following pages of that book:

Ch. 1. 172–76
Ch. 2. 42–43
Ch. 3. 106–9
Ch. 4. 179–87
Ch. 5. 208–9
Ch. 6. 96–99
Ch. 7. 52–58
Ch. 8. 143

1. THE HEART-BRAIN OF THE FOREST

1. Hoorman, "Role of Soil Fungus."
2. Wohlleben, *Hidden Life,* 86.
3. Simard, et al., "Net Transfer of Carbon."
4. Wohlleben, *Hidden Life,* 2–5.
5. Wohlleben, *Hidden Life,* 15–17.
6. Wohlleben, *Hidden Life,* 32, 249, 53; Simard, "Why Trees 'Talk'"; Frazer, "Dying Trees"; Song, "Defoliation."
7. Arvay, *Biophilia,* 9.
8. Gagliano, "Bioacoustics."
9. International, "Symbiosis."
10. Wohlleben, *Hidden Life,* 51–52.
11. Sample, "Bury the Idea"; Ephrat, "Debate."
12. Chamovitz, "What a Plant Knows."
13. Gagliano, "Green Frame."
14. Dove, "Do Plants Feel."
15. Pearl, "Asked a Biologist."
16. Gagliano, "Green Frame," 3–5.
17. Gagliano, "Green Frame," 3–5.
18. Maturana et al., *Autopoiesis,* 9.
19. Gagliano, "Green Frame."
20. Gagliano, "Green Frame"; Gagliano, *Thus Spoke,* 57–71.
21. Simard, "Why Trees 'Talk'"; Frazer, "Dying Trees."
22. Mancuso, "Roots"; Mancuso, *Revolutionary;* Mancuso and Viola, *Brilliant,* 136–44.
23. Science Daily, "Recognizing others but not yourself."
24. Wohlleben, *Hidden Life,* 82–83.
25. Wohlleben, *Hidden Life,* 243.

2. FINDING OUR PLACE IN NATURE

1. Cavalier-Smith, "Evolution"; Swimme and Berry, *Universe,* 116–25; Gavira Guerrero, *Prehistoric Life.*
2. Barnes, "What Was the Earth's First Predator and When Did it Live?"
3. University of E-N, "First Predators."
4. Barnes, "What Was the Earth's First Predator and When Did it Live?"

5. Barnes, "What Was the Earth's First Predator and When Did it Live?"
6. Graziano, "New Theory."
7. Gibson, "Evolution."
8. Stangor and Walinga, *Introduction to Psychology.*
9. Harari, *Sapiens,* 66-71; Dartnell, *Origins,* 23–24.
10. Klein, "Anatomy."
11. Fuentes, "Get the Science Right!"; Wilford, "When Humans Became Humans."
12. Harari, *Sapiens,* 1–74; Dartnell, *Origins,* 25.
13. Bird-David, "Animism"; Peoples, "Hunter."
14. Harari, *Sapiens,* 25
15. Harari, *Sapiens,* 211–12.
16. Harari, *Sapiens,* 211–12.
17. "Gaia Hypothesis."
18. Loy, *Ecodharma,* 112–13.
19. Jacquet, "Power"; Nisa and Belanger, "Can You Change."

3. HOW NATURE HEALS US

1. Qing, *Into the Forest,* 18–19.
2. Wohlleben, *Hidden Life,* 221–22.
3. Wohlleben, *Hidden Life,* 224.
4. Qing, *Into the Forest,* 58.
5. Qing, *Into the Forest,* 64.
6. Qing, *Into the Forest,* 67.
7. Qing, *Into the Forest,* 70–71.
8. Qing, *Into the Forest,* 76.
9. Qing, *Into the Forest,* 83.
10. Qing, *Into the Forest,* 87.
11. Qing, *Into the Forest,* 18, 81.
12. Qing, *Into the Forest,* 107.
13. Wohlleben, *Hidden Life,* 23; Lee and Lee, "Cardiac."
14. Williams, "Can Trees."
15. Qing, *Into the Forest,* 91.
16. Qing, *Into the Forest,* 98.
17. Qing, *Into the Forest,* 99.
18. Farrow and Washburn, "Review."

19. Walton, "Forest."
20. Twohig-Bennet and Jones, "Health"
21. Qing, *Into the Forest,* 102–3.
22. University of C-B, "Beneficial Bacteria."
23. Williams, *Nature Fix,* 130–39.
24. Williams, *Nature Fix,* 42, 43.
25. Owens, "Neanderthals"; Bradshaw, "Neanderthal Use"
26. Shurkin, "Animals."
27. Khan, "Medicinal"
28. Sumner, *Natural,* 1–20.
29. US Forest, "Active"; Elumalai and Eswariah, "Herbalism".
30. Ulrich, "View."
31. Arvay, *Biophilia,* 18.
32. Fromm, *Heart.*
33. Wilson, *Biophilia.*
34. Velux, "Global."
35. Harari, *Sapiens,* 77.
36. Harari, *Sapiens,* 83.
37. Dartnell, *Origins,* 81–82.
38. Carrington, "Humans."
39. Williams, *Nature Fix,* 46–47, 52–53.
40. Harari, *Sapiens,* 88; Harris, *Conscious,* 48–49; Pollan, *Change.*
41. Harris, *Conscious,* 50
42. Williams, *Nature Fix,* 198.
43. Worthy, "Eutierria."

4. OUR TREE CONNECTIONS

1. "Therese Neumann."
2. Russo, "Scientists."
3. Rumpho, "Solar-Powered."
4. Denison, "Evolution."
5. Bergland, "Evolutionary Biology."
6. Henrich, "Culture."
7. Melis, "Human Cooperation."
8. Robison, "People."
9. Churchland, *Conscience,* 50–54.

10. Churchland, *Conscience,* 50–54.
11. Angier, "Biology."
12. Angier, "Biology."
13. Angier, "Biology."
14. Churchland, *Conscience,* 57.
15. Lieberman, *Social;* Bergland, "Evolutionary Biology," 3.
16. Rowland, "Kindness."
17. Rowland, "Kindness."
18. Rowland, "Kindness."
19. Galante, "Effect," 1101.
20. Christakis, *Blueprint,* 13.
21. Christakis, *Blueprint,* 16.
22. Christakis, *Blueprint,* 330.
23. Christakis, *Blueprint,* 242.
24. Christakis, *Blueprint,* 318–19.
25. Stout, *Sociopath,* 35.
26. Christakis, *Blueprint,* 417–19.
27. Mancuso, *Revolutionary,* 73.
28. Mancuso, *Revolutionary,* 76–78.
29. Conradt and Roper, "Group"; Conradt, List, and Roper, "Swarm."
30. Mancuso, *Revolutionary,* 92.
31. Levy, *Collective,* 13; Jenkins, *Convergence,* 259.
32. Fisher, "Millions."
33. Christakis, *Blueprint,* 397.
34. Christakis, *Blueprint,* 402.
35. Christakis, *Blueprint,* 402.
36. Gagliano, *Thus Spoke,* 56–71; Mancuso, *Revolutionary,* 14.
37. Gagliano, *Thus Spoke,* 56–71.
38. Hance, "Plants."
39. Mancuso and Viola, *Brilliant,* 19–21, 133.
40. Mancuso and Viola, *Brilliant,* 19–20.
41. Hall, *Plants as Persons,* 147–142; Cao, Cole, and Murch, "Neurotransmitters."
42. Mancuso and Viola, *Brilliant,* 46–49.
43. Mancuso and Viola, *Brilliant,* 53.
44. Mancuso and Viola, *Brilliant,* 57–58.
45. Santora, "Do snakes have ears?"
46. "Life of an Earthworm."

47. Mancuso and Viola, *Brilliant,* 72–75.
48. Mancuso and Viola, *Brilliant,* 77–80.
49. Wohlleben, *Hidden Life,* 241–45.

5. GREENING OUR STORIES

1. Pew, "Global."
2. Gottfried, "Dominion."
3. Pope Francis, *Laudato Si';* 66.
4. Pope Francis, *Laudato Si',* 67.
5. Pope Francis, *Laudato Si',* 68.
6. Pope Francis, *Laudato Si',* 104–5.
7. Pope Francis, *Laudato Si',* 106–14.
8. Pope Francis, *Laudato Si',* 116–20
9. Pope Francis, *Laudato Si',* 137–62.
10. Pope Francis, *Laudato Si',* 215–16.
11. Pope Francis, *Laudato Si',* 218–22.
12. Berry, *Evening.*
13. The UN Environmental Programs. How Islam Can Represent a Model for Environmental Stewardship. *Reuters.*
14. Wihbey, "Green."
15. Wihbey, "Green."
16. Wihbey, "Green."
17. Environment-ecology.com/Hinduism and ecology.
18. Oxford, "Hindu Declaration."
19. Hinduism and Climate Change. http//fore.yale.edu/news/item/hindism-and-climate-change
20. Loy, *Ecodharma,* 40.
21. Kaza, *Green,* 30–31.
22. Kaza, *Green,* 76-78.
23. Loy, *Ecodharma,* 51–60.
24. Loy, *Ecodharma,* 102–6.
25. Loy, *Ecodharma,* 106–7.
26. Loy, *Ecodharma,* 112–15.
27. Loy, *Ecodharma,* 117–21.
28. Doyle, *St. Francis.*
29. Viviers, "Second Christ."

30. Giovino, *Assyrian,* 129. *World Tree Encyclopedia Britannica;* Mettinger, *Eden,* 5; *Qur'an* 14:24.
31. Proteus, *Forest,* Hageneder, *Meaning,* and "Ancient Religions."
32. Miller, "Religion."; Pollan, *Botany,* 143–45.
33. Veronese, "Psychedelic."

6. SEEING WITH A GREENER, MORE HUMBLE LENS

1. Ro, "Plant Blindness."
2. Ro, "Plant Blindness."
3. Ro, "Plant Blindness."
4. Williams, "People."
5. Williams, "People."
6. Ro, "Plant Blindness."
7. Williams, "People."
8. Wohlleben, *Hidden Life,* 224.
9. Helmenstine, "Oxygen."
10. Inglis-Arkell, "How Many."
11. "How Trees Reduce."
12. Sun, et al., "Impact."
13. Canadell, "Plants."
14. "Trees and the Water Cycle."
15. "Trees and the Water Cycle."
16. Wohlleben, *Hidden Life,* 106–7.
17. Pearce, "Rivers."
18. Parry, "How Plants Helped."
19. "Twenty-Two Benefits."
20. "Twenty-Two Benefits."
21. Dasgupta, "How Many Plants."
22. Monbiot, "*This Mess,*" 91-94.
23. Grohol and Russell, "Common."
24. Horney, *Neurotic,* 120; Horney, *New Ways,* 254–55.
25. Gazzaniga, *Ethical Brain,* 6–8.
26. Ackerman, "What Is Neuroplasticity"; Garland, "Neuroplasticity."
27. Guenther, "Climate Change."
28. Berry, *Evening,* 57.

29. Berry, *Evening,* 60.
30. Berry, *Evening,* 65; Swimme and Berry, *Universe,* 246.
31. Lewis, "Scientists."
32. Koch, "Consciousness."
33. Brogaard, "Is There Consciousness."
34. "Why Some Scientists"; Powell, "Is the Universe."
35. Skrbina, *Panpsychism.*
36. Adam Frank. 2017. "Minding Matter." Page 8 and, Harris, *Conscious,* 65–70, 81.
37. Harris, *Conscious,* 83.
38. Chopra, *Metahuman,* 118.
39. Aoki, *Entropy Principle.*
40. Leopold, *Sand County,* 201–14.
41. Harvey, "Animism."
42. Harvey, "Animism," 83.

7. RESTORING REBALANCING, REGREENING

1. Fagan and Huang, "Look at How People."
2. Berwyn, "Australia's Burning"; Toggweiler, "Shifting."
3. Slezak, "Global."
4. Irfan, "California."
5. Greenfield, "This Is Not."
6. Rodrigues, "Why Amazon."
7. Butler, "How Many Trees."
8. Macias, "America."
9. Carrington, "Tree Planting."
10. Eubanks, "How Many Trees."
11. McMahon, "Dozens."
12. Pandey, "Climate."
13. Peters, "These Drones."
14. Whiting, "This Tech Company."
15. Lant, "China."
16. Whiting, "Here's How."
17. "Regreening the Planet."
18. Greenfield, "UN Draft."
19. Zastrow, "China's Tree-Planting."

20. Greshko, "Mass Extinctions."
21. Noon, et al., "Mapping."
22. Norman, "Elon Musk."
23. Rettner, "Can You."
24. Mallonee, "Heat is On."

BIBLIOGRAPHY

Ackerman, Courtney E. "What Is Neuroplasticity? A Psychologist Explains [+14 Tools]." PositivePsychology. Last updated on November 18, 2022. https://positivepsychology.com/neuroplasticity/.

"Ancient Religions: European Tree Worship." Daily Kos. November 17, 2010. Accessed February 25, 2023.

Angier, Natalie. "The Biology Behind the Milk of Human Kindness." *The New York Times.* November 23, 2009. https://www.nytimes.com/2009/11/24/science/24angier.html.

Aoki, Ichiro. *Entropy Principle for the Development of Complex Biotic Systems.* Amsterdam: Elsevier, 2012.

Arvay, Clemens G. *The Biophilia Effect.* Louisville, Colo.: Sounds True, 2018.

Bergland, Christopher. "The Evolutionary Biology of Altruism." *Psychology Today.* December 25, 2012. Accessed February 25, 2023. https://www.psychologytoday.com/us/blog/the-athletes-way/201212/the-evolutionary-biology-altruism.

Berry, Thomas. *Evening Thoughts: Reflecting on Earth as Sacred Community.* Ed. Mary Evelyn Tucker. San Francisco, Calif.: Sierra Club Books, 2006.

Berwyn, Bob. "In Australia's Burning Forests, Signs We've Passed a Global Warming Tipping Point." Inside Climate News. January 8, 2020. Accessed February 25, 2023.

Bird-David, Nurit. "Animism Revisted: Personhood, Environment, Relational Epistemology." *Current Anthropology* 40, no. S1 (February 1999): S67–S91.

Bradshaw Foundation. "Neanderthal Use of Medicinal Plants." March 20, 2017. Accessed February 24, 2023.

Brogaard, Berit. "Is There Consciousness in Everything?" *The Superhuman Mind* (blog). *Psychology Today*. November 6, 2016. Accessed February 25, 2023.

Butler, Rhett. "How Many Trees Are Cut Down Every Year?" Mongabay. September 2, 2015. Accessed February 25, 2023.

Canadell, Pep. "Plants Absorb More CO2 than We Thought, But. . . ." The Conversation. October 14, 2014. Accessed February 25, 2023.

Cao, Jin, Ian B. Cole, and Susan J. Murch. "Neurotransmitters, Neuroregulators and Neurotoxins in the Life of Plants." *Canadian Journal of Plant Science* 86, no. 4 (October 2006): 1183-1888.

Carrington, Damian. "Humans Just 0.01% of All Life But Have Destroyed 83% of Wild Mammals—Study." *The Guardian*. May 21, 2018. Accessed February 24, 2023.

Carrington, Damian. "Tree Planting 'Has Mind-Blowing Potential' to Tackle Climate Crisis." *The Guardian*. July 4, 2019. Accessed February 25, 2023.

Cavalier-Smith, T. "Evolution and Relationships of Algae: Major Branches of the Tree of Life." *Unravelling the Algae: The Past, Present, and Future of Algal Systematics*. Edited by Juliet Brodie and Jane Lewis. Boca Raton, Fla.: CRC Press, 2007.

Chopra, Deepak. *Metahuman: Unleashing Your Infinite Potential*. New York, N.Y.: Harmony Books, 2019.

Christakis, Nicholas A. *Blueprint: The Evolutionary Origins of a Good Society*. Boston, Mass.: Little, Brown, 2019.

Churchland, Patricia. *Conscience: The Origins of Moral Intuition*. New York, N.Y.: W. W. Norton, 2019.

Conradt, Larissa, Christian List, and Timothy J. Roper. "Swarm Intelligence: When Uncertainty Meets Conflict." *American Naturalist* 182, no. 5 (2013): 592–610.

Dartnell, Lewis. *Origins: How Earth's History Shaped Human History*. New York, N.Y.: Basic Books, 2019.

Dasgupta, Shreya. "How Many Plant Species Are There in the World? Scientists Now Have an Answer." Mongabay. May 12, 2016. Accessed February 25, 2023.

Denison, R. Ford, and Katherine Muller. "The Evolution of Cooperation." *The Scientist*. January 1, 2016. Accessed February 25, 2023.

Dove, Laurie. "Do Plants Feel Pain?" *HowStuffWorks*. Accessed February 22, 2023.

Doyle, Eric. *St. Francis and the Song of Brotherhood and Sisterhood.* St. Bonaventure, N.Y.: Franciscan Institute Publications, 1996.

Elumalai, A., and M. C. Eswariah. "Herbalism: A Review." *International Journal of Phototherapy* 2 (2012): 96–105.

Eubanks, William E. "How Many Trees Are Planted Each Year?" *Green and Growing.* Accessed February 25, 2023.

Fagan, Moira, and Christine Huang. "A Look at How People Around the World View Climate Change." Pew Research Center. April 18, 2019. Accessed February 25, 2023.

Farrow, Marc R., and Kyle Washburn. "A Review of Field Experiments on the Effect of Forest Bathing on Anxiety and Heart Rate Variability." *Global Advances in Integrative Medicine and Health.* May 16, 2019.

Fisher, Laur, Robert Laubacher, and Thomas Malone. "How Millions of People Can Help Solve Climate Change." PBS Nova. January 15, 2014. Accessed February 25, 2023.

Frazer, Jennifer. "Dying Trees Can Send Food to Neighbors of Different Species." *The Artful Amoeba* (blog). *Scientific American.* May 19, 2015. Accessed February 21, 2023.

Fromm, Erich. *The Heart of Man: Its Genius for Good and Evil.* New York, N.Y.: Harper and Row, 1964.

Fuentes, Augustin. "Get the Science Right!" *Busting Myths About Human Nature* (blog). *Psychology Today.* July 17, 2017. Accessed February 23, 2023.

Gagliano, Monica. "In a Green Frame of Mind: Perspectives on the Behavioural Ecology and Cognitive Nature of Plants." *AoB Plants* 7 (2015).

Gagliano, Monica, Stefano Mancuso, and Daniel Robert. "Towards Understanding Plant Bioacoustics." *Trends in Plant Science* 17, no. 6 (2012): 323–25.

Gagliano, Monica. *Thus Spoke the Plant: A Remarkable Journey of Groundbreaking Scientific Discoveries and Personal Encounters with Plants.* Berkley, Calif.: North Atlantic Books, 2018.

"The Gaia Hypothesis." *Wikipedia.* Accessed February 23, 2023.

Galante, Julieta, Ignacio Galante, Marie-Jet Bekkers, and John Gallacher. "Effect of Kindness-Based Meditation on Health and Well-Being: A Systematic Review and Meta-Analysis." *Journal of Consulting and Clinical Psychology* 82, no. 6 (2014): 1101–14.

Garland, Eric, and Matthew Owen Howard. "Neuroplasticity, Psychosocial Genomics, and the Biopsychosocial Paradigm in the 21st Century. *Health and Social Work* 34, no. 3: (August 2009), 191–99.

Gavira Guerrero, Angeles (ed). "Prehistoric Life: The Definitive Visual History of Life on Earth," *DK Publishing*, 2012.

Gazzaniga, Michael S. *The Ethical Brain: The Science of Our Moral Dilemmas.* New York, N.Y.: Dana Press, 2005.

Gibson, Kathleen. "Evolution of Human Intelligence: The Roles of Brain Size and Mental Construction." *Brain Behavior and Evolution* 59, no. 1–2 (2002): 10–20.

Giovino, Mariana. *The Assyrian Sacred Tree: A History of Interpretations.* Fribourg, Switzerland: Academic Press Fribourg, 2007.

"Global Survey Finds We're Lacking Fresh Air and Natural Light, As We Spend Less Time in Nature." Velux Group. May 21, 2019. Accessed February 24, 2023.

Gottfried, Robert. "Dominion Over Nature and Environmental Crises-Time for Another Look." Huffpost. February 8, 2016. Accessed February 25, 2023.

Graziano, Michael. "A New Theory Explains How Consciousness Evolved." *The Atlantic.* June 6, 2016. Accessed February 23, 2023.

Greenfield, Patrick. "This Is Not How Sequoias Die. It's Supposed to Stand for Another 500 Years." *The Guardian*. January 18, 2020. Accessed February 25, 2023.

Greenfield, Patrick. "UN Draft Plan Sets 2030 Target to Avert Earth's Sixth Mass Extinction." *The Guardian.* January 13, 2020. Accessed February 25, 2023.

Greshko, Michael. "What are Mass Extinctions, and What Causes Them?" *National Geographic.* September 26, 2019. Accessed February 25, 2023.

Grohol, John M., and Tonya Russell. "Common Defense Mechanisms and Why They Work." PsychCentral. Last updated on April 28, 2022.

Guenther, Genevieve. "What Climate Change Tells Us about Being Human." *Observations* (blog). *Scientific American.* December 19, 2019. Accessed February 25, 2023.

Hageneder, Fred. *The Meaning of Trees.* San Francisco, Calif.: Chronicle Books, 2005.

Hall, Matthew. *Plants as Persons: A Philosophical Botany.* Albany, N.Y.: SUNY Press, 2011.

Hance, Jeremy. "Are Plants Intelligent? New Book Says Yes." *The Guardian.* August 4, 2015. Accessed February 25, 2023.

Harari, Yuval Noah. *Sapiens: A Brief History of Humankind.* New York, N.Y.: Harper, 2015.

Harris, Annaka. *Conscious: A Brief Guide to the Fundamental Mystery of the Mind.* New York, N.Y.: HarperCollins, 2019.

Harvey, Graham. "Animism and Ecology: Participating in the World Community." *The Ecological Citizen* 3, no. 1 (2019): 79–84.

Helmenstine, Anne Marie. "How Much Oxygen Does One Tree Produce?" *ThoughtCo.* Last updated November 19, 2019.

Henrich, Joseph and Natalie Henrich. "Culture, Evolution and the Puzzle of Human Cooperation." *Cognitive Systems Research* 7, nos. 2–3 (June 2006): 220–45.

Hoorman, James J. "Role of Soil Fungus." *Ohionline.* June 7, 2016.

Horney, Karen. *The Neurotic Personality of Our Time.* London, England: W. W. Norton, 1937.

———. *New Ways in Psychoanalysis.* London, England: W. W. Norton, 1939.

"How Trees Reduce Air Pollution." *Ecosia* (blog). April 6, 2019. Accessed February 25, 2023.

Inglis-Arkell, Esther. "How Many Plants Would You Need to Generate Oxygen for Yourself in an Airlock?" *Gizmodo.* October 26, 2012. Accessed February 25, 2023.

International Institute for Applied Systems Analysis. "Symbiosis or Capitalism? A New View of Forest Fungi." *ScienceDaily.* May 22, 2014. Accessed February 21, 2023.

Irfan, Umair. "California Has 149 Million Dead Trees Ready to Ignite Like a Matchbook." *Vox.* Last updated February 15, 2019.

Jacquet, Jennifer. "The Power of Shame Is that It Can Be Used by the Weak Against the Strong." Interview by Zoe Corbyn. *The Guardian.* March 6, 2015. Accessed February 23, 2023.

Jenkins, Henry. *Convergence Culture: Where Old and New Media Collide.* New York, N.Y.: NYU Press, 2006.

Kaza, Stephanie. *Green Buddhism: Practice and Compassionate Action in Uncertain Times.* Boulder, Colo.: Shambhala, 2019.

Khan, Haroon. "Medicinal Plants in Light of History: Recognized Therapeutic Modality." *Journal of Evidence-Based Complementary and Alternative Medicine* 19, no. 3 (July 2014): 216–19.

Klein, Richard G. "Anatomy, Behavior, and Modern Human Origins." *Journal of World Prehistory* 9, no. 2 (June 1995): 167–98.

Koch, Christof. "What is Consciousness?" *Nature.* May 9, 2018. Accessed February 25, 2023.

Lant, Karla. "China Has Officially Started Construction on the World's First 'Forest City'." *Futurism.* June 27, 2017. Accessed February 25, 2023.

Lee, Jee-Yon, and Duk-Chul Lee. "Cardiac and Pulmonary Benefits of Forest Walking Versus City Walking in Elderly Women: A Randomised, Controlled, Open-Label Trial." *European Journal of Integrative Medicine* 6, no. 1 (February 2014): 5–11.

Levy, Pierre. *Collective Intelligence: Mankind's Emerging World in Cyberspace.* New York, N.Y.: Plenum Trade, 1997.

Lewis, Tanya. "Scientists Closing in on Theory of Consciousness." *LiveScience.* July 30, 2014. Accessed February 25, 2023.

Li, Qing. *Into the Forest: How Trees Can Help You Find Health and Happiness.* New York, N.Y.: Penguin Life, 2019.

Lieberman, Matthew D. *Social: Why Our Brains are Wired to Connect.* New York, N.Y.: Crown, 2013.

"Life of an Earthworm." *JourneyNorth* (blog). Arboretum, University of Wisconsin-Madison. 2019.

Livni, Ephrat. "A Debate Over Plant Consciousness is Forcing Us to Confront the Limitations of the Human Mind." *Quartz.* June 3, 2018. Accessed February 22, 2023.

Loy, David. *Ecodharma: Buddhist Teachings for the Ecological Crisis.* Somerville, Mass.: Wisdom Publications, 2019.

Macias, Amanda. "America Has Spent $6.4 Trillion on Wars in the Middle East and Asia Since 2001, a New Study Says." *CNBC.* Last updated on November 20, 2019.

Mancuso, Stefano, and Alessandra Viola. *Brilliant Green: The Surprising History and Science of Plant Intelligence.* Translated by Joan Benham. Washington, D.C.: Island Press, 2015.

Mancuso, Stefano. "The Roots of Plant Intelligence." *TED.* October 12, 2010. YouTube video. Accessed February 23, 2023.

———. *The Revolutionary Genius of Plants: A New Understanding of Plant Intelligence and Behavior.* New York, N.Y.: Atria Books, 2018.

Maturana, Humberto R., and Francisco J. Varela. *Autopoiesis and Cognition.* Springer Dordecht, 1980.

McMahon, Jeff. "Dozens of Countries Could Reverse Carbon Emissions Just by Caring for Nature." *Forbes.* January 26, 2020. Accessed February 25, 2023.

Melis, Alicia P., and Dirk Semmann. "How is Human Cooperation Different?" *Philosophical Transactions of the Royal Society B.* September 12, 2010.

Mettinger, Tryggve N. D. *The Eden Narrative: A Literary and Religio-Historical Study of Genesis 2–3.* University Park, Pa.: Eisenbrauns, 2007.

Miller, Richard J. "Religion as a Product of Psychotropic Drug Use." *The Atlantic*. December 27, 2013. Accessed February 25, 2023.

Monbiot, George. *How Did We Get Into This Mess?* Brooklyn, N.Y.: Verso Books, 2016.

Nisa, Claudia and Jocelyn Belanger. "Can You Change for Climate Change?" *Observations* (blog). *Scientific American*. December 23, 2019. Accessed February 23, 2023.

Noon, Monica L., Allie Goldstein, Juan Carlos Ledezma, et al. "Mapping the Irrecoverable Carbon in Earth's Ecosystems. *Nature Sustainability* 5 (2022): 37–46.

Norman, Abby. "Elon Musk Tells National Governors Association How We Could Power the US with Solar." *Futurism*. July 18, 2017. Accessed February 25, 2023.

Owen, James. "Neanderthals Self-Medicated?" *National Geographic*. July 21, 2012. Accessed February 24, 2023.

Oxford Centre for Hindu Studies/Bhumi Project. "Hindu Declaration on Climate Change." 2015. Accessed February 25, 2023.

Pandey, Manish. "Climate Change: What Is Being Done Around the World to Plant Trees?" *BBC News*. September 24, 2019. Accessed February 25, 2023.

Parry, Wynne. "How Plants Helped Make the Earth Unique." *LiveScience*. February 1, 2012. Accessed February 25, 2023.

Pearce, Fred. "Rivers in the Sky: How Deforestation Is Affecting Global Water Cycles." *Yale Environment 360*. July 24, 2018. Accessed February 25, 2023.

Pearl, Mike. "We Asked a Biologist if Plants Can Feel Pain." *Vice*. Last modified September 25, 2015.

Peoples, Hervey C., Pavel Duda, and Frank W. Marlowe. "Hunter-Gatherers and the Origin of Religion." *Human Nature* 27 (May 2016): 261–82.

Peters, Adele. "These Drones Will Plant 40,000 Trees in a Month. By 2028, They'll Have Planted 1 Billion." *Fast Company*. May 15, 2020. Accessed February 25, 2023.

Pew Research Center. "The Global Religious Landscape." December 18, 2012. Accessed February 25, 2023.

Pollan, Michael. *The Botany of Desire*. New York, N.Y.: Random House, 2001.

———. *How to Change Your Mind*. New York, N. Y.: Penguin, 2018.

Pope Francis. *Laudato Si': On Care for Our Common Home*. Vatican City: Libreria Editrice Vaticana, 2015.

Porteous, Alexander. *The Forest in Folklore and Mythology*. Minneola, N.Y.: Dover Publications, 2002. First ed. 1928.

Powell, Corey S. "Is the Universe Conscious?" *NBC News*. Last updated June 16, 2017.

"Regreening the Planet Could Account for One-Third of Climate Mitigation." *Yale Environment 360*. October 17, 2017. Accessed February 25, 2023.

Rettner, Rachael. "Can You Really Make 'Meat' Out of Air?" *LiveScience*. November 18, 2019. Accessed February 25, 2023.

Ro, Christine. "Why 'Plant Blindness' Matters and What You Can Do About It." *BBC Future*. April 28, 2019. Accessed February 25, 2023.

Robison, Matthew. "Are People Naturally Inclined to Cooperate or Be Selfish?" Response by Ariel Knafo. *Scientific American*. September 1, 2014. Accessed February 25, 2023.

Rodrigues, Meghie. "Why Amazon Trees Are Especially Vulnerable to This Year's Fires." *Scientific American*. September 13, 2019. Accessed February 25, 2023.

Rowland, Lee. "Kindness—Society's Golden Chain?" *The British Psychological Society*. November 13, 2017. Accessed February 25, 2023.

Rumpho, Mary E., Elizabeth J. Summer, and James R. Manhart. "Solar-Powered Sea Slugs. Mollusc/Algal Chloroplast Symbiosis." *Plant Physiology* 123, no. 1 (May 2000): 29–38.

Russo, Karen. "Scientists Baffled by Prahlad Jani, Man Who Doesn't Eat or Drink." ABC News. May 31, 2010. Accessed February 25, 2023.

Sample, Ian. "Group of Biologists Tries to Bury the Idea that Plants Are Conscious." *The Guardian*. July 3, 2019. Accessed February 22, 2023.

Santora, Tyler. "Do snakes have ears?" *LiveScience*. May 19, 2023.

Shurkin, Joel. "Animals that Self-Medicate." *PNAS* 111, no. 49 (December 2014): 17,339–41.

Simard, Suzanne W., David A. Perry, Melanie D. Jones, David D. Myrold, Daniel M. Durall, and Randy Molina. "Net Transfer of Carbon Between Ectomycorrhizal Tree Species in the Field." *Nature* 388 (August 1997): 579–82.

Simard, Suzanne W. "Exploring How and Why Trees 'Talk' to Each Other." Interview by Diane Toomey. *Yale Environment 360*. September 1, 2016. Accessed February 21, 2023.

Skrbina, David. *Panpsychism in the West*. Cambridge, Mass.: The MIT Press, 2007.

Slezak, Michael. "'Global Deforestation Hotspot:' 3M Hectares of Australian

Forest to Be Lost in 15 Years." *The Guardian*. March 4, 2018. Accessed February 25, 2023.

Song, Yuan Yuan, Suzanne W. Simard, Allan Carroll, William W. Mohn, and Ren Sen Zeng. "Defoliation of Interior Douglas-Fir Elicits Carbon Transfer and Stress Signalling to Ponderosa Pine Neighbors Through Ectomycorrhizal Networks." *Scientific Reports* 5, no. 8495 (February 16, 2015).

Stangor, Charles, and Jennifer Walinga. *Introduction to Psychology*. Victoria, B.C.: BCcampus, 2014. Retrieved from OpenTextBC website.

Stout, Martha. *The Sociopath Next Door*. New York, N.Y.: Harmony, 2005.

Sumner, Judith. *The Natural History of Medicinal Plants*. Portland, Ore.: Timber Press, 2000.

Sun, Ying, Lianhong Gu, Robert E. Dickinson, Richard J. Norby, Stephen G. Pallardy, and Forrest M. Hoffman. "Impact of Mesophyll Diffusion on Estimated Global Land CO2 Fertilization." *PNAS* 111, no. 44 (September 2014): 15,774–79.

Swimme, Brian, and Thomas Berry. *The Universe Story: From the Primordial Flaring Forth to the Ecozoic Era*. New York, N.Y.: Harper Collins, 1992.

"Therese Neumann." *Britannica*. Last updated January 1, 2023.

Toggweiler, J. R. "Shifting Westerlies." *Science* 323, no. 5920 (March 2009): 1434–35.

"Trees and the Water Cycle." *Sustainable Footprint*. Accessed February 25, 2023.

"Twenty-Two Benefits of Trees." *TreePeople*. Accessed February 25, 2023.

Twohig-Bennett, Caoimhe and Andy Jones. "The Health Benefits of the Great Outdoors: A Systematic Review and Meta-Analysis of Greenspace Exposure and Health Outcomes." Environmental Research 166 (October 2018): 628–37.

Ulrich, Roger S. "View Through a Window May Influence Recovery from Surgery." *Science* 224, no. 4647 (April 1984): 420–21.

United States Forest Service. "Active Plant Ingredients Used for Medicinal Purposes." Accessed February 24, 2023.

University of Colorado Boulder. "Study Linking Beneficial Bacteria to Mental Health Makes Top Ten List for Brain Research." *CU Today*. January 5, 2017. Accessed February 24, 2023.

University of Erlangen-Nuremberg. "The First Predators and Their Self-Repairing Teeth." September 21, 2018. Accessed February 23, 2023.

Veronese, Keith. "The Psychedelic Cult that Thrived for Nearly 2000 Years." *Gizmodo*. February 10, 2012. Accessed February 25, 2023.

Viviers, Hendrick. "The Second Christ, St. Francis of Assisi and Ecological Consciousness." *Verbum et Ecclesia* 35, no. 1 (May 2014).

Walton, Alice G. "'Forest Bathing' Really May Be Good for Health, Study Finds." *Forbes*. July 10, 2018. Accessed February 24, 2023.

Whiting, Kate. "Here's How You Can Use Your Phone to Plant Trees." World Economic Forum. October 8, 2019. Accessed February 25, 2023.

———. "This Tech Company is Aiming to Plant 500 Billion Trees by 2060—Using Drones." World Economic Forum. December 4, 2019. Accessed February 25, 2023.

"Why Some Scientists Believe the Universe is Conscious." *Mind Matters*. August 1, 2019. Accessed February 25, 2023.

Wihbey, John. "'Green Muslims,' Eco-Islam and Evolving Climate Change Consciousness." *Yale Climate Connections*. April 11, 2012. Accessed February 25, 2023.

Williams, Florence. "Can Trees Heal People?" *TED*. June 6, 2017. Accessed February 24, 2023.

———. *The Nature Fix: Why Nature Makes Us Happier, Healthier, and More Creative*. New York, N.Y.: Norton, 2017.

Williams, Kathryn. "People Are 'Blind' to Plants and That Is Bad News for Conservation." The Conversation. September 13, 2016. Accessed February 25, 2023.

Wilson, Edward O. *Biophilia*. Cambridge, Mass.: Harvard University Press, 1984.

Wohlleben, Peter. *The Hidden Life of Trees: What They Feel, How They Communicate*. Translated by Jane Billinghurst. English edition. Vancouver, B.C., Canada: David Suzuki Institute, 2016.

Worthy, Kenneth. "Eutierria: Becoming One with Nature." *Psychology Today*. July 3, 2016. Accessed February 24, 2023.

Zastrow, Mark. "China's Tree-Planting Drive Could Falter in a Warming World." *Nature*. September 23, 2019. Accessed February 25, 2023.

INDEX

BOOKS OF RELATED INTEREST

Return of the Children of Light
Incan and Mayan Prophecies for a New World
by Judith Bluestone Polich

The Tree Angel Oracle Deck
The Ancient Path into the Sacred Grove
by Fred Hageneder
Illustrated by Anne Heng

The Secret Teachings of Plants
The Intelligence of the Heart in the Direct Perception of Nature
by Stephen Harrod Buhner

Plant Intelligence and the Imaginal Realm
Beyond the Doors of Perception into the Dreaming of Earth
by Stephen Harrod Buhner

The Tree Horoscope
Discover Your Birth-Tree and Personal Destiny
by Daniela Christine Huber

Communicating with Plants
Heart-Based Practices for Connecting with Plant Spirits
by Jen Frey
Foreword by Pam Montgomery

Plant Spirit Healing
A Guide to Working with Plant Consciousness
by Pam Montgomery

Encounters with Nature Spirits
Co-creating with the Elemental Kingdom
by R. Ogilvie Crombie